Jana Thierfelder
The Making of Scientific Knowledge

Science Studies

Jana Thierfelder, born in 1987, is a designer and anthropologist whose work bridges the arts and sciences. She has taught in the fields of ethnography, Science and Technology Studies (STS) and New Materialisms at Zürcher Hochschule der Künste and Universität Bern. She has also coordinated several interdisciplinary projects that foster dialogue and collaboration between artistic and scientific communities. Her research draws on feminist and anthropological approaches to STS, with a focus on the sensory, bodily, material and institutional conditions under which knowledge is produced in the (natural) sciences.

Jana Thierfelder

THE MAKING OF SCIENTIFIC KNOWLEDGE

Sensory and Bodily Practices
in Field Biology

[transcript]

The Open Access edition is published with the support of the Swiss National Science Foundation.

Bibliographic information published by the Deutsche Nationalbibliothek
The Deutsche Nationalbibliothek lists this publication in the Deutsche Nationalbibliografie; detailed bibliographic data are available on the Internet at https://dnb.dnb.de.

transcript Verlag | Hermannstraße 26 | D-33602 Bielefeld | live@transcript-verlag.de

Layout: Klub Galopp
Cover design: Klub Galopp
Cover illustration: Klub Galopp
Printing: Druckhaus Bechstein GmbH, Wetzlar
Proofread: Lex Academics
Copy-editing: Liz Ottosson at Better World Words
https://doi.org/10.14361/9783839435601
Print-ISBN: 978-3-8376-7931-1 | PDF-ISBN: 978-3-8394-3560-1
ISSN of series: 2703-1543 | eISSN of series: 2703-1551

Printed on permanent acid-free text paper.

Contents

Acknowledgements

Completing this work required an effort similar to that needed when hiking up a mountain, whether on boots or skis, and with or without a rope and harness. It was filled with ups and downs, fluctuating between a lack of motivation and an inability to stop. These two experiences, my academic work and hiking, provided practice for one another, as I learnt in one what I needed for the other. Both require a clear mind, mental strength, determination, and endurance. However, one does not only require strength for this endeavour, one also gains it; at times, one lacks motivation and must regain determination; at other times, one is hungry for food and must take a break to eat; or one becomes hungry for more input. Hiking and writing are iterative processes that do not always feel like uphill battles; in addition to detours, there are sections where you feel like you are going downhill before reaching the peak. However, any good mountaineer knows that mountains, particularly daunting ones, should not be climbed alone. To those who accompanied me on this adventure and supported me as best they could on my way up, I want to express my deepest gratitude.

I embarked on this climb with Michaela Schäuble at the Institute for Social Anthropology at the University of Bern, and Priska Gisler at the Institut Praktiken+Theorien der Künste at the Bern Academy of the Arts, who have accompanied me from the beginning, helped me find my way when I became lost, and guided me to ensure that I remained on track.

This endeavour would not have been possible without the open-mindedness of Michael Griesser, and that of his team, whom I met in the framework of my research, namely MA students Kate Layton-Matthews and Julian Klein, field assistants Camille Toscani and Marine Quintin, and, most recently, Miya Warrington, who has also joined Michael's project as Principal Investigator, as well as the Siberian jays I encountered. When doing fieldwork together, we were not only an ethnographer and interlocutors but also companions climbing actual hills together, which enabled me to collect empirical data and gain the insights on which this book is built, and without which none of this could have been written.

I would also like to express my deepest gratitude to Cristina Grasseni who gave me inspiring and supportive feedback on my thoughts that were developed based on her concepts.

I am particularly grateful to Irene Vögeli and Patrick Müller at the Zurich University of the Arts. I would like to thank Irene Vögeli and Hannes Rickli in particular, who introduced me to the field of art–science, helped me gain access to Michael Griesser's project, and mentored me throughout.

There were and are many more people involved in this process. I would like to mention the members of the Zeichenwerkstatt, which is run by Sigrid Adorf at the Zurich University of the Arts, my peers in the PhD colloquia at the Graduate School of the Arts at the University of Bern and the Art Academy of Bern, which I was able to join. In terms of feedback on my writing process, I want to thank Franz Neff for his advice on improving my writing. Thank you Klub Galopp for the design and transcript Verlag for turning this manuscript into a book which I can hold in my hands now.

I am grateful to my parents and my friends for their support during this process, and particularly to my companion, Sean, who accompanied me on literal mountains and nudged me forwards when I was close to giving up just below the peak. Thank you for always being there for me, encouraging me, and challenging me to conquer new peaks. Thank you, Ulysse, our latest family member, for making me look behind corners I had forgotten existed and making me discover the world with you anew, far from any disciplinary constraints.

Lastly, it was these moments in the mountains when things fell into place, thoughts came together, and concepts arose. Therefore, I cannot help but also thank the mountains for bearing with me as companions, and challenging and motivating me. They taught me what it means to persevere and helped me return to the piles of books and writing waiting for me on my desk with a fresh perspective and new ideas.

Glossary

Apparatus
Apparatus is conceptualised along the lines of physicist Karen Barad's[1] understanding, which exceeds the common understanding of apparatus as devices and instruments. Barad suggests an apparatus as a material-discursive practice that also takes its entaglement with its environment, the conditions for its production, and the epistemological context into account. This understanding resonates with Hans-Jörg Rheinberger's analysis[2] of scientific tools (or apparatuses) as things that are the result of a priori knowledge processes. Rheinberger understands these processes as the (material) stabilisation of knowledge.[3]

Design
The term 'design' creates biases, particularly in the German-speaking environment. It is often associated with the production of *designed* goods, coming from industrial design, communication design, or interface design. Design, however, when focused on the process, makes thinking-in-the-making visible, based on practice with tools and media. Thus, I propose understanding design rather as a drafting process that engages with sensory and aesthetic practices, following a formal-aesthetic approach that takes function into account. In this sense, design then could be defined as practices of drafting.

Drafting (Entwerfen)
There is no direct translation of the German term *Entwurf* into English. The closest words are 'drafting' or 'designing'. *Entwerfen* is a drafting and thinking process that requires a defined framework, and techniques, instruments (e.g. tools and media), and methods. It is defined by the use of recordings, writing and drawing systems, techniques of representation and visualisation, and a speculative and experimental procedure. The thinking can take place with, through, and in dealing with media.

Enskillment
The process of learning a specific practice and the required set of skills.

Inscription (systems)
The representations in which references circulate. It follows (within a thought collective) a pre-defined set of rules and signs to create consistency.

Intra-action
A term coined by Barad to describe the entangled nature of what is usually termed *inter*action. According to Barad, phenomena co-constitute each other only in their *intra*-actions and thus should not be looked at as separate entities (which the prefix *inter* suggests), but rather as a continuum of becoming with one another. Applied to natural scientific practices, this may mean that researcher and research object are interdependent

1
'Barad's Feminist Naturalism', *Hypatia 19*, no. 1 (2004): 142–61, https://doi.org/10.1353/hyp.2004.0012.

2
'Experimentalsysteme: Differenz, Graphematizität, Konjunktur', in *Experiment, Differenz, Schrift: Zur Geschichte Epistemischer Dinge* (Bergisch Gladbach: Basilisken-Presse, 1992), 21–32.

3
Ibid., 28.

and that one would not exist in this specific constitution without the other (as researcher and research object).

Media
Part of a technology. The form of and carriers on which recordings, references, and inscriptions are stored based on the employment of tools.

Mediation
The intra-activity of discourse, technology, research object, and the scientist (in short: apparatus) in processes of knowledge production.

Novice
Describes a person who is new to a specific work and first has to be trained and go through situated enskillment.

Objectivity
In the sciences, objectivity is a key requirement to ensure valid scientific statements and findings. Within the natural sciences, objectivity most commonly indicates that scientific truth is derived from an external referent, free from subjective biases such as emotions or a priori commitments. It emphasises the testability and reproducibility of research conducted by a neutral and impartial observer. In the humanities, however, this concept is contested.
Science historian Lorraine Daston and art historian Peter Galison[4] suggest that objectivity is always a specific form of subjectivity. It is created through a specific thought style, which defines how the scientist must act to create objectivity. Objectivity, thus, is culturally and socially determined.
Ecofeminist Donna Haraway[5] stresses that objectivity can come only from a situated position that takes the conditions of research and the potential biases of the scientists into account and makes them transparent.

Performative
This refers to a focus on the processual understanding of actions and their causal relations, for example, *science-in-the-making* as a continuous process rather than a fixed/stabilised entity.

Recording
The notation of any observations and thoughts that are transferred from an ephemeral state into a physical item and thus serve as storage, here referred to as reference.

Reference
The semiotic (often visual) representation of objects, often text or image-based in the natural sciences. Since the sensory turn of the 1980s,[6] increasing attention has been paid to senses beyond the visual – which had previously been privileged – and their representational capacities.

Referent
That which references refer to; in the sciences, this is often the research object.

Representation
The cultural practice of mirroring nature (in the case of the natural sciences) or culture (in the case of the cultural sciences), thus making it accessible to an audience. Recently, ecofeminists have been advocating for a shift from representation, which they correlate with reflection, to diffraction.[7] Representation has been accused of creating invisibilities, binaries, and affirming power relations, while a diffractive approach makes the hidden visible.

Situated/situatedness
The circumstance in which there is no *outside* perspective with universal knowledge or the capacity to produce the same.

4
Lorraine Daston and Peter Galison, *Objectivity* (New York: Zone Books, 2010).
5
Donna Haraway, 'Situated Knowledges: The Science Question in Feminism and the Privilege of Partial Perspective', *Feminist Studies 14*, no. 3 (1988): 575–99.
6
David Howes, *Sensual Relations: Engaging the Senses in Culture and Social Theory* (Ann Arbor: University of Michigan Press, 2003).
7
Karen Barad, *Meeting the Universe Halfway: Quantum Physics and the Entanglement of Matter and Meaning* (Durham, NC: Duke University Press, 2007); Donna Haraway, 'The Promises of Monsters: A Regenerative Politics for Inappropriate/d Others', in *The Haraway Reader* (New York: Routledge, 1992), 63–124.

Rather, knowledge and its production are always interdependent with the situatedness of its producers. This may be the epistemological situatedness in a discourse and the respective knowledge (see thought collective). It also encompasses institutional conditions, questions of funding, accessibility, and gender, class, and race. In short, a situated perspective takes the (hegemonic) power relations that facilitate or prevent knowledge production and may privilege certain fields over others into account. Situatedness aims at making these conditions visible and taking them into account.

Situated enskillment
The (epistemological, bodily, and sensory) enskillment of practices that are specific (situated) to a context.

Situated mediations
The entanglement between the conditions of knowledge production (as described in the term situated) and the relevant mediations.

Technology
The overarching term that also encompasses tools and media to describe the use of the relevant means and their meaning for knowledge production.

Thought collective
A term defined by Ludwik Fleck[8] to describe the social group that follows a specific style of thinking; a thought style.

Thought style
A term defined by Ludwik Fleck[9] to describe '[the readiness for] directed perception, with corresponding mental and objective assimilation of what has been so perceived. It is characterised by common features in the problems of interest to a thought collective, by the judgment which the thought collective considers evident, and by the methods which it applies as a means of cognition. The thought style may also be accompanied by a technical and literary style characteristic of the given system of knowledge'.[10]

Tool
Part of a technology. Any item that supports and facilitates the production of recordings, references, and inscriptions.

8
Ludwik Fleck, *Genesis and Development of a Scientific Fact* (Chicago: University of Chicago Press, 1979).

9
Ibid.

10
Ibid., 99.

Chapter 1: INTRODUCTION

This book is based on five years of fieldwork undertaken with a team of evolutionary biologists studying Siberian jay birds *(Perisoreus infaustus)* in northern Sweden. Ethnographic data were gathered during two extended field trips to the team study sites in the boreal forests of the Sápmi region, and several preparatory and follow-up visits to the researchers' offices. Employing a grounded theory approach, my findings were informed by participant observation and analysis of the scientists' field materials – some collected in the field and others preserved in the biologists' archives. The ethnography provides new perspectives on scientific knowledge production by investigating the role that tools, design decisions, and representational practices play in the scientific research process and how they are developed to produce knowledge. I do not aim to improve scientific practices or change the way data are presented in scientific literature. Instead, I aim to elucidate, for a non-scientific audience, the practices involved in the production of scientific papers.

In my attempt to do so, I combine two field sciences: anthropology and evolutionary biology. While the case study focuses on a group of evolutionary biologists studying the behaviour of Siberian jays, a bird species, the methodological approach derives from anthropology. Combining anthropology with science and technology studies (STS) allows for the strengthening of an ecofeminist perspective on knowledge production, attending specifically to the role of the human and non-human actors.

The merger between STS and anthropology is informed by a design perspective. Visualisation practices first caught my attention as a formally trained visual communication designer and anthropologist. I became aware of the gap in the publication of scientific research by studying the raw field notes and data of evolutionary biologists. While field notes usually remain hidden in archives, I suggest that they are, to the same extent as the graphs in published articles, the result of specific design decisions that follow formal-aesthetic principles.

Accordingly, the way in which visualisation is employed in the field sciences suggests some commonalities with design, including the frequently used visual systems of organising, structuring, arranging, and categorising information. The practices of visualisation and the resulting material are highlighted by including design in the discourse. Thus, a design-informed perspective is valuable in reimagining knowledge production because it addresses the usually invisible aspects of scientific research, such as sensory and bodily skills, creativity, emotionality, aesthetics, and implicit thinking.

Evolutionary biology, like most branches of biology, is very generally characterised by positivism, objectivity, and empirical measurement. In contrast, anthropology is associated with constructivism and the knowledge contingency that entails. However, successfully responding to real-world exigencies may involve practices that fall somewhere between the two. Thus, combining these two disciplines in an experimental setting may transcend the boundary between the so-called *hard* and *soft* sciences in favour of new approaches to scientific knowledge production in the field sciences. This perspective extends beyond positivist and constructivist practices of worldmaking to practices that produce 'faithful accounts of a "real" world'[1] by revealing the ontological and epistemological entanglements in which scientists find themselves during knowledge production.

Bearing this in mind, I aim to highlight the so-called *softness* of the natural sciences, an aspect that exists prior to the transformation of knowledge into scientific fact through systemic thinking, natural laws, and model-based reasoning. I also aim to address the practice–theory divide that manifests in natural sciences publications where most of the practical processes, observations, and manipulations that lead to conclusions are excluded from the discussion. This exclusion becomes clear when examining visualisation practices that extend beyond technologies of vision to other senses, as I shall demonstrate. This ethnography combines the *hard* and *soft* sciences into a productive discourse to reveal what is epistemologically and ontologically concealed. The biologist and ecofeminist Donna Haraway notes:

> It matters what matters we use to think other matters with; it matters what stories we tell to tell other stories with; it matters what knots knot knots, what thoughts think thoughts, what descriptions describe descriptions, what ties tie ties. It matters what stories make worlds, what worlds make stories.[2]

In this text, I expand on her musings by asking: what practices make worlds and what stories do those practices tell? The worlds I discuss in this ethnography are created through scientific knowledge production. By adopting a pragmatic approach, I question which sensory and bodily practices constitute data collection and processing, enable thoughts, and shape thinking. Ultimately, this monograph is focused on the epistemologies and ontologies of thinking and doing in the field sciences.

1 Haraway, 'Situated Knowledges', 579.

2 Donna Haraway, *Staying with the Trouble: Making Kin in the Chthulucene* (Durham, NC: Duke University Press, 2016), 12.

One notable feature of natural science journals – particularly those in the life sciences – is the rare visibility of actual research *objects*, such as birds, mammals, or fish, as they exist in the natural world. Readers are typically presented with image complexes[3] presenting data in charts, numbers, figures, tables, and graphs embedded within a larger framework of information. The data are usually presented in a highly formalised and abstract way, dominant in the natural sciences. Scholars perpetuate a certain perception of their discipline by adhering to this format. While presenting results in this way may make perfect sense to a natural scientist, publications often become completely detached from their referent (in nature) by obscuring the events and people who contributed to the scientific process. Important practices that lead to scientific advancements become naturalised, ahistorical, and inaccessible to the wider public. In addition, the biologists themselves, and the research objects and tools used, become almost entirely invisible, as if they were never part of the process of knowledge production. Condensed methodology sections remain that obscure the individual stages of knowledge production through intense filtering of the research data.[4] In this sense, scientific journals become a platform where scientists perform what Haraway has dubbed the 'god trick',[5] the phenomenon that creates the

3
Martina Merz, 'Bildkomplexe als Geschichten: Naturwissenschaftler erzählen', in *Erzählen in den Wissenschaften,* 2009.

4
When it comes to methodology in the natural sciences, many publications focus solely on research methods. However, these are usually highly technical, excluding, e.g., the specifics of documentation or sensory attunement. They address a peer-to-peer audience and do not, as I aim to do with my science and technology studies (STS) approach, attempt to facilitate communication between the sciences and the public. Thus, these publications are less accessible to those outside the relevant disciplines.

5
Donna Haraway uses the metaphor of the 'god trick' in her critique as one in which objectivity is created through 'a view from above' (589): a 'god trick of seeing everything from nowhere' (581). However, according to her, 'that view of infinite vision is an illusion' (582) and the result of the 'highest technoscientific visualizations' (584). She argues that 'the view from a body, always a complex, contradictory, structuring, and structured body, versus the view from above, from nowhere, from simplicity' (589) is situated and does not try to hide the bodily and sensory involvement. She proposes situated knowledges as an alternative to objectivity. Haraway, 'Situated Knowledges'.

gap between the presentation of the research object and research problem, and the scientific conclusions. This text connects design, social anthropology (here referred to as anthropology), and evolutionary biology to bridge this gap in the discourse.

At its core, this presentational gap is characterised by the disappearance of the 'long series of manipulations' described by anthropologist and philosopher Bruno Latour,[6] which results in a contradiction in the scientific apparatus of representation. On the one hand, according to Latour, '[T]he more steps there are in between the objects and those who make judgments about them, the more robust those judgments will be' (ibid.). In this sense, the chain of manipulation is a distinguishing feature of *good* scientific work. On the other hand, when it comes to the publication of results, '[The scientists] suddenly are more than happy to display one isolated image extracted out of the chains as "the definitive proof"[7] of the phenomenon they wish to describe'.[8] What the scientific community considers a sign of excellence is invisible to the public. The information that is ultimately published in scientific journals is limited to whatever lies on either side of the gap, and scientific facts have been directly extracted from the scientific object without any steps in between.

To expand on this observation, I focus on what has been eliminated during the process. Revealing the scientific practices obscured in the presentation of the data provides valuable insights, both epistemological and ontological, into knowledge production. By examining this gap, I highlight existing aspects of the scientific process and reintroduce them to the discourse of knowledge production in the

6
Bruno Latour, 'The More Manipulations the Better', in *New Representation in Scientific Practice,* ed. Catelijne Coopmans et al. (Cambridge, MA: MIT Press, 2014), 348.

7
While for public representation, the rhetoric of the images and the written account often come across as if they were 'definitive proof', as Latour refers to it, I would, instead, call it an 'approximation to truth' with reference to my interlocutors that I shall introduce in the following sections of the text.

8
'The More Manipulations the Better', 348.

natural sciences. In this way, a different narrative can be constructed; one that makes the matters, events, thoughts, descriptions, and connections that create worlds visible again. In this book, I bridge the gap by bringing the biologists, the birds, and their environment back into the discourse, thereby revealing the processes that produce scientific knowledge.

The following section introduces the case study on which the ethnography is based: a team of evolutionary field biologists working in northern Sweden with Siberian jays *(Perisoreus infaustus),* a bird species uniquely associated with the Indigenous Sámi people who live in that area.

Case Study

Siberian jays *(Perisoreus infaustus)* are found in the Sápmi region of northern Sweden and share a deep cultural connection with the Sámi people. The birds are often described as 'charismatic' and 'almost tame,'[9] tending to appear near people and viewed as 'hunters' friends' in the northern hemisphere. Siberian jays often appear during food offerings – a Sámi tradition in which food is placed in trees – not only seeking nourishment but also seemingly drawn to human presence.[10] At the same time, they have traditionally been regarded as birds of 'ill omen'. The Latin word 'infaustus' means unlucky or associated with bad luck or even death. Most importantly for research purposes, Siberian jays are easy to study because they are a curious rather than shy species, making them readily observable.

9
Ingela Bergman and Lars Östlund, 'A Sacred Tree in the Boreal Forest: A Narrative about a Sámi Shaman, Her Tree, and the Forest Landscape', *Human Ecology* 50, no. 6 (December 2022): 1023–33, https://doi.org/10.1007/s10745-022-00365-x.

10
Ibid.

PD Dr Michael Griesser[11] has served as the principal investigator (PI) of the project since 2004, and in 2023, Dr Miya Warrington joined him as co-PI. The researchers in Michael's team aim to discover why animals cooperate with each other and live in family groups, and to learn more about communication and language from animals. The team use field experiments and behavioural, longitudinal, and comparative data to explore mechanisms underlying sociality. These insights are relevant for conservation, species resilience, and climate change adaptation and mitigation.[12] The team works with Siberian jays because the birds' social system is unusual. Besides a breeding pair, 'family' groups of jays may include their own offspring that remain with their parents for up to four years, and unrelated non-breeders. This variation in kinship among non-breeders allows the scientists to study the benefits of family living.[13] They investigate the *proximate* and *ultimate* causes of cooperation,[14] a common focus in behavioural biology: the cost and benefit of certain behaviours and how birds (or other animals) cooperate within their social structures. For this, they work not only in laboratories and offices but also in the field, where an extensive part of data collection takes place. Siberian jays are widespread across northern Eurasia. Thus, the biologists' study site is in northern Sweden, in Swedish Sápmi, the land of

11
Hereafter, I shall refer to Michael by his first name, as I do with the other biologists I observed in the field. I introduce those who are officially involved in the study by their full name, position, and affiliation. I refer to the students and researchers who took part in the study only temporarily by their first names.

12
Michael Griesser: C-Wild Griesser. Retrieved from: https://sites.google.com/view/c-wild-griesser. 28/4/2025.

13
Ibid.

14
During experiments, biologists are interested in studying the behaviours of birds in terms of their cost (investment) and benefit (advantages). With cost–benefit analyses in behavioural biology, the biologists address the investment (cost) against the benefit (advantage) of survival or breeding advantages. In such analyses, the consequences of behaviour are addressed on two levels, answering the 'why' and 'how' of the proximate and ultimate causes. The immediate, proximate causes motivate behaviour, such as a warning call for a predator that allows other birds to escape. The long-term functions and ultimate causes relate to the survival of the genetic information of a breeding couple when protecting and raising their offspring.

Europe's only Indigenous people, the Sámi. However, the study site is located there not only because of the research focus but also because of the history of the Siberian Jay Project, which was initiated by the teacher Folke Lindgren, who lived there. Carl von Linné also travelled to the region during his eighteenth-century research journeys to gather ecological and anthropological knowledge, as documented in his posthumously published research journals.[15] Here, biologists observe the birds' behavioural responses to their experimental settings. To turn these observations and experiments into scientific data, the biologists must document them. Using visualisation practices, they create permanent inscriptions[16] that are processed until the final images are produced. However, they must also find, attract, register, and study the birds to collect data. For this, particularly during fieldwork, the biologists engage in various sensory and bodily practices, and they use several visual tools to record, store, and transport their observations.

The biologists' field notebooks containing handwritten notes, drawings, protocols, and datasheets, in addition to hard drives containing video recordings and other raw data, provide interesting material for anthropological STS. While final scientific images are usually created digitally, hiding the processes involved in the raw field data, field notebooks reveal the human engagement and practices of knowledge production from start to finish.

The use of tools and media such as notebooks and pencils to capture data has a long tradition, particularly in evolutionary biology. Early naturalists such as Maria Sibylla Merian, Alexander von Humboldt, Carl von Linné, and Alfred Russel Wallace (another early proponent of natural selection,

15
Nelson Goodman, *Ways of Worldmaking* (Brighton: Harvester Press, 1978).
16
Bruno Latour, 'The "Topofil" of Boa Vista: A Photo-Philosophical Montage', in *Pandora's Hope: Essays on the Reality of Science Studies,* ed. Bruno Latour (Cambridge, MA: Harvard University Press, 1999), 24–79; Latour, 'The More Manipulations the Better'; Bruno Latour, 'Visualisation and Cognition: Drawing Things Together', in *Knowledge and Society: Studies in the Sociology of Culture Past and Present: A Research Annual, Vol. 5, 1984,* ed. Henrika Kuklich and Elizabeth Long (Greenwich, CT: JAI Press, 1984), 1–40.

along with Charles Darwin, who should also be mentioned here) made use of handwritten field notes. Their early field notebooks reveal that evolutionary biology observations always involve a combination of noting, collecting, describing, and categorising – along with relevant bodily and sensory practices – to create inscriptions. In addition, I suggest that biologists engage in design practices.

Charles Darwin's archives were filled with images that had been produced in cooperation with artists such as John and Elizabeth Gould, William Swainson, Joseph Wolf, Oscar Gustav Rejlander, Briton Rivière, and Thomas Woolner.[17] However, arts and science were not so easily separable, especially in the early nineteenth century. The Goulds, for instance, were considered not only artists but also naturalists. Together, they contributed to ornithology through their widely recognised survey, *The Birds of Australia,* with Elizabeth serving as a skilled illustrator and John as an obsessive bird collector.[18] The archives of these early scientists, which include scientific image production, have recently attracted the attention of art and science historians.[19] These scholars have observed that the archives would be less comprehensive and many of the resulting scientific insights may not have been possible without the contributions of artists and their visualisation practices. Their goal is also to understand how forms of knowledge and modes of production came together in early scientific studies. These were the first collaborations between the arts and sciences; current collaborations between the two

17
cf. Julia Voss, *Darwins Bilder: Ansichten Der Evolutionstheorie 1837–1874* (Berlin: Fischer Taschenbuch Verlag, 2007), 332.

18
John Gould, *The Birds of Australia* (London: Richard and John E. Taylor, 1848).

19
Lorraine Daston and Peter Galison, 'The Image of Objectivity', *Representations* 40, no. 1 (1992): 81–128; Peter Galison, 'Objectivity Is Romantic', in *Humanities and the Sciences,* ed. Jerome Friedman, Peter Galison, and Susan Haack (ACLS, 2000), 15–43; Christoph Hoffmann and Alexandre Métraux, 'Working with Instruments: Ernst Mach as Material Epistemologist, a Short Introduction', *Science in Context* 29, no. 4 (2016): 429–33; Christoph Hoffmann and Barbara Wittmann, 'Introduction: Knowledge in the Making: Drawing and Writing as Research Techniques', *Science in Context* 26, no. 2 (2013): 203–13.

fields often have different intentions.[20] However, sensory and bodily engagements beyond visualisation practices have received little attention thus far. To understand these worldmaking[21] practices beyond the analysis of visual material in the field sciences, I qualitatively observe and analyse a case study in evolutionary field biology in the chapters that follow.

Overview of Chapters

The following chapters continue to lay the foundation for the study. Chapter 2 describes the methods used, focusing on the dual aspect of my ethnographic fieldwork observing the evolutionary biologists, and the biologists' fieldwork collecting data from the Siberian jays. Chapter 3 presents the theoretical framework on which the work is based, drawing mainly on STS and anthropology and further informed by design. Key concepts are introduced in that chapter, which are essential for understanding the substantive chapters that follow. Chapters 4–6 represent the research process and build on one another, focusing on the practices of scientific fieldwork.

Chapter 4 discusses the practices of *Preparing* that the biologists and I engaged with prior to the data collection. This chapter describes the requirements for scientists to be part of the study and focuses on preparation, from arriving at

20
Art programmes in scientific institutions and laboratories, as well as transdisciplinary fine arts–science collaborations, have become popular in the past few decades. However, they appear to serve the purpose of scientific communication through the fine arts to increase public interest in the sciences, which are often accused of functioning in ivory towers. Actual fine arts–science collaborations in which the arts are involved in the scientific process – from funding applications to production of the results – appear to be rare. However, a growing publication record of recent scientific papers resulting from such collaborations has been observed, substantiating the epistemological role of fine arts beyond science communication (e.g. Amber Dance, 'Art Graft: Putting an "A" into "STEMM"', *Nature* 590 (2021): 351–53; Amanda C. Niehaus, 'Tell the Stories in Your Science', *Nature* 557 (2018): 269; Matthias C. Rillig and Karine Bonneval. 'The Artist Who Co-Authored a Paper and Expanded My Professional Network', *Nature*, 27 February (2020): 1–8.

21
Goodman, *Ways of Worldmaking.*

the field camp and preparing for field days to a more detailed analysis of the skills that must be acquired beyond formal training in peer-learning settings within the field. These include wayfinding and registering, which I refer to as *situated enskillment*. I conclude by juxtaposing these preparations with my own entry into the field as an anthropologist and my enskillment as a participant observer.

Chapter 5, *Collecting*, provides a detailed description of how the biologists collect data on the birds. These practices are shaped by approaching the birds in their territories and identifying and observing them. I pay particular attention to the sensory alignment necessary to find, attract, and observe the birds. I conclude my detailed account of situated mediations by revealing the entanglements of sensory, bodily, and technological practices. My reflection on the relationship between anthropology and the biological research practice focuses on *participant behaviour observation*, a speculative compound suggested as a way to reflect on the human–non-human interactions at stake during my observations as an anthropologist and during the biologists' observations of the birds.

Chapter 6 focuses on the processing of data based on practices of *Producing*. In this section, I focus on the transformation of the raw field data into final datasets and graphs based on technological operations. The metaphor of filtering is introduced and used to describe what is filtered out in three stages. Along with this, the locations of practice change from field to office, and the research becomes less dependent on the field and, as I shall argue, less situated as well. I conclude with a *thick description by means of visualisation*, which aims to emphasise the difference in data handling between anthropology, where thick description is part of qualitative data, and biology, where everything that is *thick* must be filtered out to produce universally valid models of nature. The final chapter, Chapter 7, provides some concluding thoughts and insights for the future.

A Note on Images as Interludes

The individual chapters in this book are, where relevant, introduced by image-interludes, a concept inspired by the creation of mood boards in design. The images brought together at the beginning of the chapters visually set the tone and juxtapose the dismantling that, as I shall argue throughout the text, occurs during scientific knowledge production. Rather than putting them between the text, these image collections serve as an independent sensory narrative with images that simultaneously form part of my data.

The images are employed diffractively to thicken my written account and create transparency, not merely by illustrating what is already there but rather by extending my empirical descriptions. Studying the images that are marked as 'Figures', thus becoming part of a referential system in the text, the reader will notice that they do not always correlate with what is written. Rather, they may notice small shifts, for example, observing a different researcher doing the practices described, thus adding to the narration beyond illustrating what has been said. This approach helps to exemplify the repetitiveness of the fieldwork and add an additional perspective by showing different researchers based on different modalities. This should not be misunderstood as an affirmation of scientific reproducibility and objectivity, suggesting that individual researchers do not have an impact on the data collection. This is not my aim. Rather, I want to bring together several layers describing my observations and take the opportunity to not merely duplicate information but extend it.

From the biologists' perspective, as I shall show, much of what is made visible in the interludes is treated as an aesthetic surplus and will become a waste product. This also becomes visible in the decreasing number of images with every step of research. The interludes offer a space for this surplus. They make it visible as raw data that, from my perspective, are a source of insights into the conditions of knowledge production.

Chapter 2: METHODS

This book attempts to trace the scientific process from the initial preparations for a field trip for data collection to the processing of the data in university offices, and the development of models of the world that are created as facts through this process. In this way, I weave together material from my empirical work, which combines data from my fieldwork, with the biologists' archives that I was given access to, and my theoretical framework. I aim to highlight that which is often invisible in scientific research by following a rigorous empirical approach. I position this work in the methodological framework of anthropology and ethnography, discussing how humans make sense of the world through scientific work. I do so by attending to the not-yet-stable – the work-in-progress – and thus the processual, practical, and performative, rather than by primarily attending to stabilised scientific knowledge. This chapter sets out to provide an overview of the research design and its philosophical underpinnings, along with details of the procedures, data collection and analysis processes that took place.

Two research projects are combined in this study: the biologists' study of the social behaviour of the birds, which serves as the case study, and my ethnographic study of the biologists. Whereas the biologists use methods that are considered *exact* (in a scientific sense) to study and document the behaviour of the birds, I use *soft*, qualitative methods in my ethnographic study. I address the sensory aspect from two perspectives: on the one hand, it is an ethnography about the role of the senses in biological research; on the other, I conduct my research through sensory attunement,[1] which will become visible in the empirical accounts of this study.

I study the visual traces and the raw field material of the research process. The analysis considers how such traces and materials help to categorise and structure information, taking account of the aspects that are filtered out from raw data to publication, and the transformations the visualisations undergo to arrive at the final images and represent new knowledge. The biologists and the birds are especially visible in the raw field material of the biological study; however, for publication they must be obscured. What is left are inscriptions that refer to them. In this book, I bridge the gap by bringing the biologists, the birds, and their environment back into the discourse, thereby revealing the processes that produce scientific knowledge.

Following the method of a 'patchy Anthropocene',[2] I combine disciplines in an experimental way for a critical analysis of the dynamics between the human and non-human, adding complexity rather than reducing it. In this way, I attempt to interweave the social sciences with the natural sciences to overcome dichotomies and divides. Understanding this ethnography as 'patchy' allows for 'attending to specificity without being parochial'.[3] It highlights the openness of the research and its experimental approach, as well as its incompleteness of knowing. Rather than providing

1
Sarah Pink, 'Doing Sensory Ethnography', in *Doing Sensory Ethnography*, 2009, 7–23, https://doi.org/10.4135/9781446249383.n2.

2
Anna Lowenhaupt Tsing, Andrew S. Mathews, and Nils Bubandt, 'Patchy Anthropocene: Landscape Structure, Multispecies History, and the Retooling of Anthropology: An Introduction to Supplement 20', *Current Anthropology* 60, no. S20 (2019): S186–97, https://doi.org/10.1086/703391.

3
Ibid., 187.

a seemingly exhaustive narrative 'from above', I reveal different perspectives by collecting (rather than hunting for)[4] 'patches' of stories that facilitate a reimagining of the sciences by taking in a partial perspective from below.[5] 'From below' opposes the so-called 'god's perspective' from above. Haraway associates the 'from above' view with a *neutral* scientist, a *modes witness* that quietly and invisibly observes nature from outside. However, a perspective from below also yields observations and creates different power relations that question rather than affirm hierarchies. 'From below' positioning is not outside the world it aims to study but rather situated within it; it is earthbound. Acting from below also means thinking from an entangled and involved perspective, becoming part of the research, and taking this *being part of,* and thus also being able to take on response-ability[6] seriously.

By attending to the body and senses, this ethnography explores how we might change the perspective in the sciences from a seemingly uninvolved practice 'from the outside' towards one of physical and sensory engagement, by collecting stories about the environment and bringing them together. By making these stories accessible, this endeavour also addresses questions concerning the ontological gap between nature and culture, between human and non-human, and between subjects and objects.[7]

4
Ursula K. Le Guin, 'The Carrier Bag Theory of Fiction', *The Ecocriticism Reader: Landmarks in Literary Ecology,* 1996, 149–54.

5
Cf. Haraway, 'Situated Knowledges'.

6
Haraway, *Staying with the Trouble.* In Haraway's discussion of the concept of the 'god trick', she continues that one is response-able only from a situated perspective that is grounded within the world rather than above. This plays on the duality of the meaning of being responsible for the stories that create worlds (in this case, the scientific results), but also having the ability to respond, thus creating a dialogue rather than a monologue 'from above'. Ultimately, response-ability is a question of ethics, whereas ontologies refer to questions of being, and epistemologies to questions of knowing. Situated knowledges aim to bring these together as 'ethico-onto-epistem-ology' (cf. Barad, *Meeting the Universe Halfway,* 90) instead of treating them as separate entities.

7
I refer to the term 'object' to describe the Siberian jays as the research *objects* of my case study. However, I argue that the boundary between the Siberian jays as *objects* and *subjects* is fluid and not always clearcut. Hence, where suitable, I also use the term 'research subject' to refer to the birds.

Before I elaborate further on my methodological approach and framework, I discuss the perception that anthropology 'borrows its tradition of modern fieldwork from biology (Stocking 1953) and developed from this its own robust methods for studying human worlds'.[8] This view indicates commonalities between the methods of data collection in field biology and anthropology that deserve attention.

In this case, both evolutionary biology and ethnography can be considered field sciences. Similar to design, there are thus parallels between the scientific processes I describe. I do not aim to juxtapose the two, but rather to use them to reflect on my own practices. These moments of self-reflection appear as *patches* throughout the text, mostly at the end of the chapters. As these parallels became apparent to me only during the analysis of my data, I did not collect data on my own methods while collecting data on the biologists. Instead, the patches are retrospective reflections drawn from the analysis that reveal how insights gained from studying knowledge production in evolutionary biology can contribute to social anthropological methodology.

This text draws on my fieldwork experiences, guided by grounded theory and built on participant observations,[9] as well as the analysis of the biologists' field materials gathered either directly during fieldwork or by studying their archives. Over the past five years, I accompanied Michael and his group of evolutionary biologists in their scientific process, collecting my ethnographic data during two field trips (in the spring of 2015 and 2020) to their study area in Sweden, complemented by several prepara-

8
George W. Stocking, 'The Ethnographer's Magic: Fieldwork in British Anthropology from Tylor to Malinowski', in *Observers Observed: Essays on Ethnographic Fieldwork,* 1985. As cited in: Nils Bubandt, Astrid Oberborbeck Andersen, and Rachel Cypher, 'Introduction. Rubber Boots Methods: Outline for a Multispecies Study of the Anthropocene', in *Rubber Boots Methods for the Anthropocene,* ed. Nils Bubandt, Astrid Oberborbeck Andersen, and Rachel Cypher, Doing Fieldwork in Multispecies Worlds (University of Minnesota Press, 2022), 1–36, http://www.jstor.org/stable/10.5749/j.ctv2h43983.4.

9
Kathy Charmaz, *Constructing Grounded Theory: A Practical Guide through Qualitative Analysis* (London: SAGE, 2006); Danny L. Jorgensen, *Participant Observation: A Methodology for Human Studies* (Thousand Oaks, CA: Sage Publications, 1989); Karen O'Reilly, *Ethnographic Methods* (New York: Routledge, 2005), https://doi.org/10.4324/9780203864722.

tory and follow-up visits to their offices. Some of the most important encounters were informal conversations during laboratory visits and online meetings that focused on the progress of their research. I dedicated most of my ethnographic attention to the sensory and bodily practices of the biologists during the production of scientific output. I focus particularly on how the employment of visual tools, such as binoculars, manual note-taking, and video cameras, influences the scientific practices of observation, data collection, and knowledge production.

I shadowed one biologist each day during the two field studies, recording my observations in my field notebook and supplementing them with photographs. The biologists attracted the birds and observed them, took blood samples, gathered body data, ringed them, and conducted experiments; they recorded observations on the birds' *proximate and ultimate causes of cooperation* (cf. footnote 14 in Chapter 1) using partially formalised procedures (protocols, notebooks, sound and video recordings, side-notes, sketches, and maps) to document their data. The role of tools, media, and the formal-aesthetic decisions made throughout the process of knowledge production about birds is not of interest to the researchers and receives little to no epistemological attention. The sensory practices and bodily knowledge that accompany data collection, and thus the research process, in the natural sciences is neither reflected nor formalised. This knowledge is filtered out and disappears from the study discourse and findings. Examining data-collection practices makes it possible to study how cognitive and epistemic aspects of sensory and bodily engagement, along with their material results such as sensory capacities, notation techniques, and visual skills, contribute to scientific knowledge production and maintain their visibility.

I spent much of my time actively participating in the biologists' fieldwork, which allowed me to understand the individual steps of their research and follow the practices that produced scientific facts. Imitating the biologists and assisting them as a pseudo-biologist allowed me to follow their thought processes and gain a more profound understanding of their work. As part of my participant observation during their daily fieldwork, I gained insight into their drafting techniques and – mainly visual – data-collection practices in the field. Insights were gained into the acquisition of data-collection skills; how these skills are used differently by the biologists depending on skill, experience, and habit; and how their drafting techniques affect their fieldwork practices.

This approach took place alongside conversations and qualitative interviews that often occurred during fieldwork, particularly when waiting for the birds and navigating the study sites. Living together in the field station – and thus witnessing each step of the fieldwork routines, which involved preparation and reflection in the mornings and evenings – and sharing spare time provided many opportunities to exchange and gain further information. In addition to this, I spent much of my time listening to the interlocutors – the biologists – arguing, planning, and discussing their fieldwork. This provided insight into how the biologists decide on a method, their individual perspectives on them, and how they work with different kinds of visual data.

As the biologists mainly work independently in the study areas, I was able to accompany each of them for several days on both of my field trips with Michael and his team. This allowed me to gain insights into their individual working styles, and their knowledge and familiarity with fieldwork and their profession in general. I noticed differences and grew to understand the nature of formalised and predetermined practices, in contrast with practices that take place at the biologists' discretion. In addition, I gained insights into how new biologists learn to conduct fieldwork for the first time. During the data evaluation in the field camp, I observed how the team collaboratively thought, communicated, and worked based on their raw data and organisational materials. I learnt about the drafting methods employed from data collection to publication, how scientists process and digitalise their datasets, and what role different kinds of images and image production play, for instance, as representational images, operative images[10] or mnemonics. The following section introduces the evolutionary biologists' archived dataset, which I was fortunate to be given access to, and which provided a rich source of data for the study.

Archive

To follow these filtering processes and translational steps, I refer to a dataset and archive provided by the biologists in my case study. The evolutionary biologists working with Michael on Siberian jays draw on data that date back to the 1950s. Back then, Swedish school teacher and ornithologist, Folke Lindgren, began studying Siberian jays in his garden behind his house in Arvidsjaur, expanding his study site over time as birds started to disappear here because of wood logging. From 1952 to 1988, Lindgren gathered data on the birds, later monitoring up to 15 bird territories containing one bird *family* each. Lindgren was the first person to become curious about the unusual family life of these birds.

My dataset comprises more than 50 notebooks, several folders with datasheets, hard drives with video material, and digital records. These complement the field data I collected in notebooks and diaries, and through audio and video recordings. I compiled excerpts from this archive to trace the relevant

10
Martina Merz and Inge Hinterwaldner, 'Neue Bilder, Modelle und Simulationen: Zwischen Repräsentativität und Produktivität', in *Handbuch Wissenschaftssoziologie,* ed. Sabine Maasen et al. (Wiesbaden: Springer, 2012), 303–16.

processes and to identify the role of individual practices beyond the obvious. Thus far, this archive has been used only for biological analysis. Bringing this material into the field of design- and anthropology-informed STS research can provide valuable insight into scientific knowledge production from a practice-based perspective and make the process of scientific knowledge production visible.

By doing this, I expand on theories on drafting and writing in scientific knowledge production[11] by focusing on their practical aspects. Moreover, I do not refer to historical material but rather to research practices as they are currently performed in behavioural studies in evolutionary biology. Last, the use of visualisation has been widely reflected in anthropology and the social sciences,[12] allowing for its juxtaposition with the way in which visualisation and scientists themselves have been reflected in the natural sciences. Therefore, this expands the methodological discourse in the natural sciences by focusing on the specific practices and bodily, sensory, and implicit knowledge associated with scientific methods.

11
See, e.g., Nicolas Gansterer, *Drawing a Hypothesis: Figures of Thought*, ed. Gerald Bast (Wien/New York: Springer, 2011); Hoffmann and Wittmann, 'Introduction: Knowledge in the Making: Drawing and Writing as Research Techniques'; Karin Krauthausen, 'Vom Nutzen des Notierens, Verfahren des Entwurfs', in *Notieren, Skizzieren. Schreiben und Zeichnen als Verfahren des Entwurfs.*, ed. Karin Krauthausen and Omar W. Nasim (Zürich/Berlin: diaphenes, 2010), 7–26; Karin Krauthausen et al., *Notieren, Skizzieren. Schreiben und Zeichnen als Verfahren des Entwurfs*, ed. Karin Krauthausen and Omar W. Nasim (Zürich/Berlin: diaphanes, 2010); Omar W. Nasim, 'Extending the Gaze: The Temporality of Astronomical Paperwork', *Science in Context* 26, no. 2 (2013): 247–77, https://doi.org/10.1017/S0269889713000057.

12
See, e.g., Gemma Anderson, 'On Drawing as a Way of Knowing', in *Drawing as a Way of Knowing in Art and Science* (Bristol/Chicago: Intellect, 2017), 3–27; Andrew Causey, *Drawn to See: Drawings as an Ethnographic Method* (Toronto: University of Toronto Press, 2017); Emma K. Frow, 'Drawing a Line: Setting Guidelines for Digital Image Processing in Scientific Journal Articles', *Social Studies of Science* 42, no. 3 (2012): 369–92; Haidy Geismar, 'Drawing It Out', *Visual Anthropology Review* 30, no. 2 (2014): 97–113; Anna Grimshaw and Amanda Ravetz, 'Drawing with a Camera? Ethnographic Film and Transformative Anthropology', *Journal of the Royal Anthropological Institute* 21, no. 2 (2015): 255–75; Michael Guggenheim, 'The Media of Sociology: Tight or Loose Translations?', *British Journal of Sociology* 66, no. 2 (2015): 345–72; Hoffmann and Wittmann, 'Introduction: Knowledge in the Making: Drawing and Writing as Research Techniques'; Claudia Mitchell et al., 'Picturing Research', in *Picturing Research* (Rotterdam: SensePublishers, 2011), 1–16, https://doi.org/10.1007/978-94-6091-596-3_1; Michael Taussig, 'What Do Drawings Want?', *Culture, Theory and Critique* 50, no. 2–3 (2009): 263–74; Michael Taussig, *I Swear I Saw This: Drawings in Fieldwork Notebooks, Namely My Own* (Chicago: University of Chicago Press, 2011).

A critical reading of the (raw) data raises questions about how scientific knowledge acquisition is characterised primarily by visual practices. In this regard, I enquire how 'processes of classification',[13] such as the visual structuring and organisation of data, influence thinking and the acquisition of knowledge. Thus, the previously mentioned gap between the visual components of raw images and their public representation can be observed and analysed. Although the drawings and notes in the field notebooks of the collaborating biologists still appear personal, intuitive, subjective, and observant – that is, close to everyday life – these data are replaced by a scientific semiotics that represents rationality, objectivity, and reproducibility.

In addition, the scientists evolve from being in a situated practice in which they are visible in the raw data as individuals – for instance, in their handwriting – to an objectification of themselves where all traces of them are eliminated. Moreover, the images of the scientists evolve from practitioners in the field to intellectuals in offices, suggesting that the practices during fieldwork and the intellectual processing thereof can be separated. Further study of the different kinds of visual language in the biologists' field notes will provide insights into the involvement of certain practices in the construction of a scientific representation of nature. In short, I follow the material traces of the biologists to reimagine scientific work.

The following chapter provides an overview of the theoretical framework informing this study, allowing readers to combine their understanding of what took place in the field and offices of the evolutionary biologists with how the events and observations were understood.

13
Charles Goodwin, 'The Blackness of Black: Color Categories as Situated Practice', in *Discourse, Tools and Reasoning: Essays on Situated Cognition,* ed. Lauren B. Resnick et al. (New York: Springer, 1997), 111–40.

Chapter 3: THEORETICAL FRAMEWORK

The ethnography presented in this text draws on two main areas of study – STS and anthropology – complemented by a design-informed perspective. The combination of the two fields enables an empirical examination of the practices of preparing, collecting, and producing knowledge in the field sciences. It provides the theoretical framework for empirical and sensory practices, the role of embodiment and skill, and the social and institutional conditions of scientific knowledge production. Against the background of design, the combination of these fields allows technology to be considered, with special consideration of their cognitive, epistemic, and methodological agencies.

I highlight all the practices in which the biologists in my case study engage, from data collection to publication, thereby drawing on the commonalities between ethnographic accounts and the work of field biology, which are 'quite similar' given that they 'involve[s] watching and describing the social relations the analyst finds',[1] and share certain

1
Anna Tsing, 'When the Things We Study Respond to Each Other: Tools for Unpacking "the Material"', in *Anthropos and the Material*, ed. Penny Harvey, Christian Krohn-Hansen, and Knut G. Nustad (Duke University Press, 2019), 228.

historical commonalities. By creating these relationships, I extend anthropological scholarship to the sciences and vice versa, thus productively blurring the boundaries, as I believe that it is in these boundaries that the unknown can be found. With this, I seek to contribute a fresh perspective on what it means to produce scientific knowledge, as a practice that appreciates that it is 'earthbound'[2] and 'situated'.[3] The goal is to ethnographically focus on the onto-epistemologies[4] of science-in-the-making as sensory, situated, and mediated practices in which the knowing subject and the object of knowledge production 'intra-actively'[5] become entangled. This chapter provides further detail on these key theoretical frameworks, their commonalities, and key concepts that are then interwoven throughout successive chapters. But first, let us consider the context and rationale for the theoretical framework adopted in this work.

Context and Rationale

Over the past few decades, advances in molecular and nanoscience – and the scientific observations they have enabled – have drawn increasing attention from scholars in the sciences, technology studies, and the history and philosophy

2
Bruno Latour, *Down to Earth: Politics in the New Climatic Regime* (Cambridge, UK: Polity, 2018).

3
Haraway, 'Situated Knowledges'.

4
'[...] knowing is a material practice of engagement as part of the world in its differential becoming intra-acting' (Barad, *Meeting the Universe Halfway*, 89). Thus, epistemologies and ontologies are inseparable. Barad would add ethics to intra-dependency of the concepts, as represented in this neologism.

5
Barad, *Meeting the Universe Halfway*.

of science.[6] Numerous studies have been published on subjects such as CERN (The European Organization for Nuclear Research) and nanoscale image production. However, the excitement about these scientific achievements has overshadowed interest in the field sciences, such as evolutionary biology or ecology, which have received little attention.[7]

However, with the alarming consequences of climate change becoming increasingly visible in recent years, the relationship between human beings and nature has been questioned anew, and, as a result, the role of the sciences has shifted.[8] We currently face the sixth mass extinction, through irreversible change to ecosystems and the permanent loss of flora and fauna, which has returned the public's attention to the natural world. With this development, scientific attention has also returned to evolutionary biology.

6
See, e.g., Barad, *Meeting the Universe Halfway;* Anne Dippel, 'The Big Data Game: On the Ludic Constitution of the Collaborative Production of Knowledge in High-Energy Physics at CERN', *Naturwissenschaften, Technik und Medizin* 25, no. 4 (2017): 485–517; Judith Dobler, 'Collaborative Imaging: The Communicative Practice of Hand Sketching in Experimental Physics', in *Proceedings of DRS 2016: Future-Focused Thinking, Vol. 3, Sec. 5 Aesthetics, Cosmopolitics and Design* (Design Research Society, 2016), 997–1011; Judith Dobler, 'Drawing Together: Collaborative Design Practices in Experimental Physics', in *Nordes: Design+Power* (Nordes, 2017); Hoffmann and Wittmann, 'Introduction: Knowledge in the Making: Drawing and Writing as Research Techniques'; K. Knorr-Cetina, *Epistemic Cultures: How the Sciences Make Knowledge* (Cambridge, MA: Harvard University Press, 1999); Bruno Latour and Steve Woolgar, *Laboratory Life: The Construction of Scientific Facts,* ed. Jonas Salk (Princeton, NJ: Princeton University Press, 1979); Merz and Hinterwaldner, 'Neue Bilder, Modelle und Simulationen: Zwischen Repräsentativität und Produktivität'; Hans-Jörg Rheinberger, *Toward a History of Epistemic Things: Synthesizing Proteins in the Test Tube* (Stanford, CA: Stanford University Press, 1997), 325.

7
See, e.g., Michael G. Hadfield and Donna J. Haraway, 'The Tree Snail Manifesto', *Current Anthropology* 60, no. 20 (2019): S209–35; Latour, 'The "Topofil" of Boa Vista: A Photo-Philosophical Montage'.

8
Adam Frank, Marcelo Gleiser, and Evan Thompson, *The Blind Spot: Why Science Cannot Ignore Human Experience* (Cambridge, MA: The MIT Press, 2024).

Against this background, philosophical discourse on nature, ecology, and evolution has received increasing attention.[9] Moreover, in anthropology, nature and its entanglements with the social and the political realm have become the subject of recent scholarly interest.[10] What remains missing from the literature is an STS perspective that both philosophically and culturally situates ecology within the discourse and examines the practices of knowledge production that shape current philosophical and anthropological discourse.

Introduction to Key Concepts

The primary theoretical framework of this text draws on the feminist concept of objectivity and the conditions under which scientific knowledge is produced. Two of the most important concepts in this research are Haraway's situated

9 See, e.g., Emanuele Coccia, *The Life of Plants: A Metaphysics of Mixture* (Cambridge: Polity, 2019); Vinciane Despret, *What Would Animals Say If We Asked the Right Questions?* (Minneapolis: University of Minnesota Press, 2016); Michael Marder, 'For a Phytocentrism to Come', *Environmental Philosophy* 11, no. 2 (2014): 237–52, https://doi.org/10.5840/envirophil20145110; Michael Marder, *Plant-Thinking: A Philosophy of Vegetal Life* (New York: Columbia University Press, 2013); Timothy Morton, *Dark Ecology: For a Logic of Future Coexistence* (New York: Columbia University Press, 2016); Merlin Sheldrake, *Entangled Life: How Fungi Make Our Worlds, Change Our Minds, and Shape Our Futures* (London: The Bodley Head, 2020); Thom Van Dooren, *Flight Ways: Life and Loss at the Edge of Extinction,* Critical Perspectives on Animals: Theory, Culture, Science, and Law (New York: Columbia University Press, 2016).

10 See, e.g., Wendy Harding, 'Anna Tsing, Heather Swanson, Elaine Gan, Nils Bubandt (Eds.), Arts of Living on a Damaged Planet: Ghosts and Monsters of the Anthropocene', *Miranda,* no. 16 (2018): 0–5, https://doi.org/10.4000/miranda.11648; Robin Wall Kimmerer, *Braiding Sweetgrass: Indigenous Wisdom, Scientific Knowledge and the Teachings of Plants*. Minneapolis: Milkweed Editions, 2013; Eduardo Kohn, *How Forests Think: Toward an Anthropology Beyond the Human* (Los Angeles: University of California Press, 2013); Anna Lowenhaupt Tsing, *The Mushroom at the End of the World: On the Possibility of Life in Capitalist Ruins* (Princeton: Princeton University Press, 2015); Rachel Mundy, 'Birdsong and the Image of Evolution', *Society and Animals* 17, no. 3 (2009): 206–23, https://doi.org/10.1163/156853009X445389; Andrew Whitehouse, 'Listening to Birds in the Anthropocene: The Anxious Semiotics of Sound in a Human-Dominated World', *Environmental Humanities* 6 (2015): 53–71.

knowledges[11] and physicist and philosopher Karen Barad's concepts of agential realism, such as intra-activity and onto-epistemologies.[12]

Regarding the role of visual skill in scientific practices of worldmaking, I refer to anthropologist Cristina Grasseni's theories of skilled visions and skilled mediations.[13] I attempt to relate them to what skill may mean in terms of situatedness, embodiment, and sensory practices as part of the apparatus of knowledge production, extending them to what I call *situated mediations*.

Additional concepts woven throughout the monograph are circulating reference and immutable mobiles,[14] as formulated by Latour, based on one of the few STS studies in the field sciences. Together with pedologists in Boa Vista in the Amazon rainforest, he studied how scientists *translate* their research subject, the soil of the rainforest, into their laboratories across the world, and how they develop papers based on these data.[15]

While Latour speaks of cascades of inscriptions, transformation, transfers, and translational steps, I prefer the concept of filtering to discuss what is filtered out, lost, and left behind, as well as what is gained through these steps of transformation in the research process.

11
Haraway, 'Situated Knowledges'.

12
Barad, *Meeting the Universe Halfway*.

13
Cristina Grasseni, 'Skilled Vision: An Apprenticeship in Breeding Aesthetics', *Social Anthropology* 12, no. 1 (2004): 41–55; Cristina Grasseni 'Skilled Visions: Toward an Ecology of Visual Inscriptions', in *Made to Be Seen: Perspectives on the History of Visual Anthropology*, ed. Marcus Banks and Jay Ruby (Chicago: University of Chicago Press, 2007), 19–44; Cristina Grasseni *Skilled Visions: Between Apprenticeship and Standards*, ed. Christina Grasseni (New York, Oxford: Berghahn Books, 2009); Cristina Grasseni, 'Beauty as Skill and "Common Sensing"', in *Anthropology and Beauty: From Aesthetics to Creativity*, ed. Stephanie Bunn (London: Routledge, Taylor & Francis Group, 2018); Cristina Grasseni and Thorsten Gieser, 'Introduction: Skilled Mediations', *Social Anthropology* 27, no. 1 (2019): 6–16.

14
Latour, 'The More Manipulations the Better'; Latour and Woolgar, *Laboratory Life*.

15
Latour, 'The "Topofil" of Boa Vista: A Photo-Philosophical Montage'.

Objectivity

In their monograph *Objectivity,* Lorraine Daston and Peter Galison[16] connect approaches from the history of science to the history of art by relating visualisation practice to knowledge production. To them, the construction of objectivity is intertwined with the subjectivity of the researchers[17] and the 'moralization of objectivity'.[18] From their perspective, the researching subjects must first familiarise themselves with the practices of scientific work; in this sense, it is also a matter of belonging to a thought collective.

Daston and Galison describe this history of objectivity in relation to image production in the sciences, showing how it is shaped by historical developments. In this context, they emphasise the 'diversity and contingency of the components that make up the current concept of objectivity'.[19] By studying the historical development of scientific images, the authors claim that it was not technological development alone that shaped these images, but rather the prevailing zeitgeist and virtues associated with each era. Thus, they focus instead on the subjectification practices required by researchers to produce objective images.

This perspective also allows the authors to take institutional, political, and scientific developments into account and assess how they affect authorship and the production of evidence in their visual representation. This subjectification is the result of the scientists' attunement to a certain style, and to rules, modes, and practices as a prerequisite for scientific image production.

16
Daston and Galison, *Objectivity*.

17
Daston and Galison, 'The Image of Objectivity', 82.

18
Ibid.

19
Ibid.

Achieving objectivity in this sense is entangled with the so-called 'epistemic virtues'.[20] These virtues, namely 'truth to nature',[21] 'mechanical objectivity',[22] and 'trained judgement',[23] subjectivise the scientists by means of visual practices in such a way that the resulting images are considered objective. In addition to this, Daston and Galison develop three ideal personas for scientists:

> The sage, whose well-stocked memory synthesizes a lifetime of experience with skeletons or crystals or seashells into the type of that class of objects; the indefatigable worker, whose strong will turns inward on itself to subdue the self into a passively registering machine; the intuitive expert, who depends on unconscious judgment to organize experience into patterns in the very act of perception.[24]

Daston and Galison's contribution helps to transform the concept of objectivity into one of subjectivity, overcoming the predominant notion of a neutral observer, similar to a god's view from nowhere, as suggested by Haraway. I combine this with Barad's perspective on science as an intra-active entanglement between knowing subjects and objects of knowledge production. These co-constitute each other intra-actively in scientific practices, thus forming apparatuses. This perspective allows one to address the subject–object dichotomy in conventional science production and question the apparent separation of the two. Objectivity, in particular, as the term itself suggests, creates a subject–object divide between the researching subject, that is, the evolutionary biologist and the object of research, the Siberian jays.

20
Daston and Galison, *Objectivity*, 39.
21
Ibid., 55.
22
Ibid., 115.
23
Ibid., 309.
24
Ibid., 44.

While a subject–object divide can be observed in scientific knowledge production, a separation between humans and non-humans is also at stake. During fieldwork, I observed biologists becoming birds, aligning with the environment, engaging with the birds and forest on a sensory level, and even addressing the birds by imitating bird sounds, thus attempting to meet them on equal terms. The biologists entered the field as human beings, but during fieldwork they increasingly aligned their movements, behaviours, and even their thinking with that of the birds. Therefore, fieldwork is highly shaped by complex sensory engagement and multimodal social processes, and does not allow one to maintain these dualities. Accordingly, while an ontological separation of nature and culture still occurs in the publication of scientific information in the natural sciences, these ontologies collapse during the production process, and human and non-human lifeworlds overlap as naturecultures,[25] challenging the dominant *great divides*.

Situated Knowledges

Haraway developed the concept of situated knowledges in collaboration with other feminist scholars, including Sarah Harding, Nancy Hartsock, and Karen Barad, famously coining the term in her 1988 essay, *Situated Knowledges: The Science Question in Feminism and the Privilege of Partial Perspective.* In this paper she suggested a feminist rereading of the, often problematic, notion of objectivity. Situated knowledges help to overcome the objectivity–relativism duality. While objectivity suggests neutrality, universality, and impartiality, relativism renders knowledge a question of opinion, 'a way of being nowhere while claiming to be everywhere equally'.[26] In Haraway's words: 'The moral

25
Donna Haraway, *When Species Meet* (Minneapolis: University of Minnesota Press, 2008).
26
Haraway, 'Situated Knowledges', 584.

is simple: only partial perspective promises objective vision'.[27] Therefore, it is not so much about doing away with objectivity but reconceptualising it.

Haraway employs situated knowledges as a concept that preserves claims to objectivity but reveals the partiality of each perspective. She considers the possibility of theories for the *real* world, based on 'strong truth claims' (Young, 2022) that are not solely socially constructed and contingent. Thus, situated knowledge makes the political, institutional, epistemological, ethical, and ontological circumstances of knowledge production visible, situating it within the power relations that enable them. Feminist situated knowledge considers the diverse material-semiotic agencies involved in the production of knowledge. They can be understood as active instruments that produce knowledge, an apparatus of bodily production creating 'material-discursive entanglements' in which epistemologies and ontologies become interlaced as onto-epistemologies.[28]

Situated knowledges are a 'view from somewhere'[29] as opposed to the 'god trick', which is a 'gaze from nowhere', an 'immaterial gaze', and an impartial perspective that has the capacity 'to represent while escaping representation'.[30] Yet the 'power to see', to actively organise the world, is shaped by 'semiotic-material'[31] conditions that are far from self-evident. They are technologically, socially, and psychologically learnt.

Haraway continues with reference to technologies that work based on vision, a metaphor that is widely used in feminist science studies as vision that is characterised by distance rather than entanglement. Ecofeminists, in particular,

27
Ibid., 583.
28
Barad, *Meeting the Universe Halfway*.
29
Haraway, 'Situated Knowledges', 590.
30
Ibid., 581.
31
Ibid., 585.

aim to deconstruct this notion of vision as objective owing to its distance, emphasising instead that *seeing* in science is an embodied practice shaped through apparatuses, manipulations, and bodily engagement. It is also a matter of touching and, thus, not sensory engagement from a distance but based on proximity:

> The 'eyes' made available in modern technological sciences shatter any idea of passive vision; these prosthetic devices show us that all eyes, including our own organic ones, are active perceptual systems, building on translations and specific ways of seeing, that is, ways of life. There is no unmediated photograph or passive camera obscura in scientific accounts of bodies and machines; there are only highly specific visual possibilities, each with a wonderfully detailed, active, partial way of organizing worlds.[32]

Vision, as Haraway describes it, is not a neutral and objective activity performed by the eye; instead, it is skilled, trained, and active. Guided and directed by the viewer's position and situatedness within the world, it is also based on disciplinary education, knowledge, skill, and interest: it is skilled vision. In relation to scientific research, it is not a 'view from above, from nowhere, from simplicity', but rather a 'view from the body',[33] which is situated, partial, and thus not universal.

Skilled Visions / Skilled Mediations

Vision, 'a semiotic eye',[34] is not merely looking; it is intertwined with an 'apprenticeship of the eye'[35] – a process of training and knowledge in the employment of this bodily tech-

32
Ibid., 583.
33
Ibid., 589.
34
Grasseni, 'Conference on Neuroesthetics'.
35
Grasseni, 'Skilled Vision: An Apprenticeship in Breeding Aesthetics', 42.

nique. Skilled visions are 'invisible, embodied, sensorial, tacit, a result of incogation, training and acculturation'[36] as well as a 'complex relation between attention, habit, and representational capacities'.[37] Skilled visions can be described as a professional visual competence based on the biologists knowing what to look for and how to look at it when doing fieldwork or analysing scientific images. Thus, biologists mediate between the thought collective and the object of interest.

Cristina Grasseni and Thorsten Gieser developed the concept of skilled mediations as a response to the sensory turn in the social sciences, considering vision as a result of bodily and sensory practice, apprenticeship, and skill, which often occur in combination with technologies. Thus, they extended Grasseni's 'skilled visions'.[38] These authors aim 'to advance the epistemological understanding within anthropology of mediation, a concept indicating the technical and sensory apprenticeship that is intrinsic to enskillment – a process that we identify as crucial to ethnographic practice and anthropological understandings'.[39]

While skilled visions lacked the role of mediation through 'focusing media', with skilled mediations the authors focus on 'how fieldwork experience is itself crucially mediated by tools, educated attention and relevant media'.[40] Thus, skilled mediation is culturally determined and produced by 'situated learning',[41] which defines how the world is studied. Mediation is also always a question of the thought

36
Grasseni, 'Conference on Neuroesthetics'.

37
Grasseni, 'Beauty as Skill and "Common Sensing"', 224.

38
Grasseni, 'Skilled Vision: An Apprenticeship in Breeding Aesthetics'; Grasseni, 'Skilled Visions: Toward an Ecology of Visual Inscriptions'; Grasseni, 'Conference on Neuroesthetics'.

39
Grasseni and Gieser, 'Introduction: Skilled Mediations', 7.

40
Ibid., 7.

41
Ibid., 8.

collective, which defines a certain perspective that is shaped by discourse. It is informed by a thought style and is a question of bodily production.

Situated knowledge should be viewed in conjunction with skilled mediations because it allows for an expansion of Gieser and Grasseni's concept through its onto-epistemological account as material-discursive practices of bodily production, which are also always situated. This allows for emphasis on the entanglements between vision and the senses, and the technologies, epistemologies, and ontologies of visual and sensory attunement. Therefore, I consider skilled mediations part of situated mediations.

Representations are produced based on cultural and bodily techniques.[42] From these perspectives, social anthropologists contribute an additional analytical framework to the role of image production as a method with epistemological value. Here, the discussion of 'anthropology as a form of image-making practice'[43] is widespread regarding 'ways of knowing, skilled practice, improvisation and the imagination',[44] as well as 'tacit'[45] and 'embodied knowledge'.[46] With regard to embodied knowledge, Ingold[47] and Taussig's[48] theories of drawing have provided intensive reflection on 'its significance as a special kind of knowledge practice'[49]

42
Marcel Mauss, 'Techniques of the Body', in *Journal de Psychologie Normal et Patholigique,* ed. Marcel Mauss (London: Routledge, 1935), 271–93; Harun Maye, 'Was Ist Eine Kulturtechnik?', *Zeitschrift Für Medien und Kulturforschung,* no. Heft 1 (2010): 121–35.

43
Grimshaw and Ravetz, 'Drawing with a Camera?', 255.

44
Ibid., 215.

45
Michael Polanyi and Amartya Sen, *The Tacit Dimension* (Chicago: University of Chicago Press, 2006).

46
Causey, *Drawn to See;* Tim Ingold, *Lines – A Brief History* (New York: Routledge, 2007); Almut-Barbara Renger et al., 'Körperwissen: Transfer und Innovation', *Paragrana* 25, no. 1 (2016): 13–19.

47
Ingold *Lines – A Brief History;* Tim Ingold, *Making: Anthropology, Archaeology, Art and Architecture* (New York: Routledge, 2013); Tim Ingold, *The Life of Lines* (New York: Routledge, 2015).

48
Taussig, 'What Do Drawings Want?'; Taussig, *I Swear I Saw This.*

49
Grimshaw and Ravetz, 'Drawing with a Camera?', 255.

and an analysis of the interfaces between drafting and thinking. Thus, these accounts allow for a better understanding of how drafting practices influence sensory engagement with the environment.

Immutable Mobiles

Scientific knowledge production, like the production of objectivity, is always the result of researchers' practical engagement with their epistemic object.[50] Steve Woolgar and Bruno Latour[51] expand on the role of practices within the sciences in their laboratory studies. They focused on the 'social construction of scientific facts', which is also the subtitle of their 1979 published work, thereby adopting a social constructivist perspective that assumes that scientific work is possible only through a network of (also social) practices that extend beyond the scientific process. They focus particularly on the process of writing and the importance of the result – the scientific paper constructed through writing practices including transcriptions, inscriptions, and translations.

These steps can also be applied to visual practices and scientific image production, and can be understood as 'chains of transformation'[52] or 'a regulated series of transformations, transmutations, and translations'[53] based on inscriptions. Inscriptions are 'a small window through which one could read a very few signs from a rather poor repertoire (diagrams, blots, bands, columns)'.[54] Thus, these materials result from scientific operations performed on material carriers, such as notebooks. Latour continues: 'All these

50
Rheinberger, *Toward a History of Epistemic Things.*

51
Latour and Woolgar, *Laboratory Life.*

52
Latour, 'The "Topofil" of Boa Vista: A Photo-Philosophical Montage', 70.

53
Ibid., 58.

54
Latour and Woolgar, *Laboratory Life*, 4.

inscriptions, as I called them, [are] combinable, superimposable and could, with only a minimum of cleaning up, be integrated as figures in the text of the articles people were writing'.[55] They work as circulating references that keep the scientific process steady.

Scientific certainty can be produced only through these chains of signs that, according to these questions, become stronger over the course of their referential production, with one sign leading to the next to return them to the discourse[56] or, rather, turn them into scientific objects. Signs, then, result from the translation processes by means of representational practices that must be kept steady as references but also mobile, as they circulate simultaneously. Throughout this book, I attend to the translational practices that these signs undergo, from reference to the bird to representation in publication, discussing them as filters.

Adopting this perspective allows one to draw attention to the agency of signs, such as references produced through representational practices that enable consistency in data collection. In this regard, researchers have studied the correlation between visual and material practices in scientific processes, and the associated translation steps that lead from the researched object to a scientific fact.[57]

The final section of this chapter considers design research and the perspectives it offers in a transdisciplinary study such as the present one.

55
Ibid.

56
Bruno Latour, 'Circulating Reference: Sampling the Soil in the Amazon Forest', in *Pandora's Hope: Essays on the Reality of Science Studies* (Cambridge, MA: Harvard University Press, 1999), 32.

57
See, e.g., Stephan Kammer et al., *Spuren Erzeugen. Zeichnen und Schreiben als Verfahren der Selbstaufzeichnung,* ed. Barbara Wittmann (Zürich/Berlin: diaphanes, 2009); Knorr-Cetina, *Epistemic Cultures;* Krauthausen et al., *Notieren, Skizzieren. Schreiben und Zeichnen als Verfahren des Entwurfs;* Bruno Latour, 'Drawing Things Together. Die Macht der Unveränderlich Mobilen Elemente', in *ANThology: Ein Einführendes Handbuch Zur Akteur-Netzwerk-Theorie,* ed. Andréa Belliger and David J. Krieger (Zürich/Berlin: transcript Verlag, 2006), 259–307; Merz and Hinterwaldner, 'Neue Bilder, Modelle und Simulationen: Zwischen Repräsentativität und Produktivität'.

Design Research

Following the STS example of primarily adopting the natural sciences as objects of social scientific analysis, design research uses the natural sciences to address questions about the epistemologies of design. The approach provides new perspectives on the practices of scientific work, from data collection in the field to the production of scientific papers. Since the design methods movement of the 1960s, design researchers have been working on definitions and conceptualisations of design methods and processes, with special regard to their cognitive-epistemic and communicative-interactive role.[58] These researchers pursue the goal of establishing design as an independent academic field and exploring the transdisciplinary interfaces between design and knowledge.[59]

Within this context, drafting a design is understood as a human activity[60] and a problem-solving endeavour.[61] Thus, design is based on embodied and experiential knowledge and unconscious, implicit decisions. For this reason, it permeates all areas of life, including science. Nonetheless, design researchers Claudia Mareis and Christof Windgätter stated that 'the styles of thought, action and publication prevalent in the scientific world hardly developed a

58
See, e.g., Nigel Cross, 'Design Research: A Disciplined Conversation', *Design Issues* 15, no. 2 (1999): 5–10; Nigel Cross, 'Designerly Ways of Knowing: Design Discipline Versus Design Science', *Design Issues* 17, no. 3 (2001): 49–55, https://doi.org/10.1162/074793601750357196; Christopher Frayling, 'Research in Art and Design', *Royal College of Art Research Papers* 1, no. 1 (1993): 1–5; Donald A. Schön, *The Reflective Practitioner: How Professionals Think in Action*, ed. Donald A. Schön (New York: Basic Books, 1983); Herbert A. Simon, *The Sciences of the Artificial*, ed. Herbert A. Simon, vol. 33 (Cambridge, MA: MIT Press, 1997).

59
See, e.g., Claudia Mareis, *Design als Wissenskultur: Interferenzen zwischen Design-und Wissensdiskursen seit 1960* (Bielefeld: Transcript, 2011); Claudia Mareis et al., *Wild Thing. Unordentliche Prozesse in Design und Wissenschaft*, ed. Claudia Mareis and Christof Windgätter (Berlin: Kulturverlag Kadmos, 2019).

60
Hans Kaspar Hugentobler et al., *Designwissenschaft und Designforschung: Ein Einführender Überblick* (Luzern: Hochschule Luzern, 2010), 11.

61
Annette Diefenthaler, 'Problemsolving', in *Design Dictionary. Perspectives on Design Terminology*, ed. Michael Erlhoff and Tim Marshall (Basel: Birkhäuser, 2007), 307.

relationship to design themes'.[62] They add, 'There is still a hierarchical-chronological difference [...] between the actual generation of knowledge and its subsequent presentation'.[63] In other words, even though scientific research is always associated with questions of design, this interface has mostly been disregarded.

Therefore, the relationship between the production of knowledge and representational activities during research calls for a renegotiation of the boundaries between design and science. Judith Dobler's PhD project *Drawing Together*[64] is, to the author's knowledge, the only known example of an interface that addresses both design practices, such as drawing techniques and forms of knowledge, media practices, and collaborative activities. Dobler's ongoing empirical study at the University of Potsdam focuses on physicists and their drawing processes, thereby also addressing the epistemological gap by introducing the field of STS, ethnography, and design to the discourse.

In the substantive data chapters that follow, the practices of preparing (Chapter 4), collecting (Chapter 5), and producing (Chapter 6) knowledge will be described in detail and used to elucidate the hidden processes and epistemological gaps in the evolutionary biology case study.

62
Cf. Mareis et al., *Wild Thing. Unordentliche Prozesse in Design und Wissenschaft,* 10.
63
Ibid.
64
Judith Dobler, '"Flatten the Curve". The Performative Embodiment of Image, Speech, and Gestures in Scientific Communication from a Design Research Perspective', in NERD – New Experimental Research in Design 2 (Basel: Birkhäuser, 2021), 24–35, https://www.degruyterbrill.com/document/doi/10.1515/9783035623666-003/html; Judith Dobler, 'Collaborative Drawing as Knowledge Practice', in Approaches to Drawing in Architectural and Urban Design, ed. F. Colonnese, N. Grancho, and R. Schaeverbeke (Newcastle upon Tyne: Cambridge Scholars Publishing, 2024), 489–504.

Chapter 4: PREPARING

On 6 March 2015, I meet Michael at the Zurich airport. We had finalised our preparations for fieldwork during the past few weeks. We take three flights to Arvidsjaur, stopping in Amsterdam and Stockholm. While Michael feels entirely at ease on this journey, which he has completed at least twice a year over the past 20 years, I am slightly nervous and uncertain about what to expect in northern Sweden. For Michael, it is a routine part of his research, but I have never travelled this route before.

Once we drop off our luggage, go through security, and get onto the plane, we take our seats next to each other. We start talking and Michael shares his general perspective on science. I make my first entries in my field notebook.[1] Michael shares that being a scientist does not mean that he wants to prove anything; instead, he wants to develop approximations to truth claims about the birds and their environment, and describe them as accurately as possible.[2] Michael learnt only some of his methods during his formal

1
Chapters 4–6 are written mainly in the present tense to reflect their foundation in contemporaneous field notes and to convey an authentic, immediate account of events, experiences, and perceptions as they unfolded during ethnographic observation.

2
I later saw a similar idea expressed in a film about evolutionary biologist Lynn Margulis, where the public's perception of natural scientific practice as offering definitive *proof* of natural phenomena, was depicted in the same way. The film suggested instead that scientific practice should be understood in Michael's sense. Cf. John Feldman (2017). Symbiotic Earth: How Lynn Margulis Rocked the Boat and Started a Scientific Revolution [Video]. Vimeo. https://vimeo.com/ondemand/symbioticearthhv.

education at university. Much of his practical knowledge was acquired before university through bird censuses and bird watching. Most of his skills in observing and experimenting were acquired in the field by accompanying others or as part of his own research or by teaching himself. In particular, he learnt methods for the Siberian Jay Project from Dr Jan Ekman, a biologist based at the University of Stockholm, Sweden, who joined the project in the 1980s. Michael then developed those methods further.

In Arvidsjaur, Michael studies the birds' behaviour. For Siberian jays to successfully adapt to their habitat and survive, the most important requirements are access to food, protection from predators, and opportunities for reproduction. In evolutionary biology, these three elements can be subsumed under the term 'behaviour'. Consequently, the birds' strategies to meet these needs and ultimately survive are the biologists' focus, and they refer to the birds' ability to adapt as their 'fitness'. The birds must maintain their fitness to survive – their primary focus throughout their lifespan. Their behaviours depend on their social and ecological environments, and their familiarity with other birds. Eventually, they are shaped by and depend on evolution, as their behaviours shape their chances of survival and the reproduction of their genes.[3]

Lindgren, the first to observe these birds in Arvidsjaur, defined how the generations after him would observe the birds and collect data. In a paper Michael co-authored with Sarah Lagerberg, entitled *Long-Term Effects of Forest Management on Territory Occupancy and Breeding Success of an Open-Nesting Boreal Bird Species, the Siberian Jay,* they state:

> We determined territory occupancy and reproductive success on the territories with the help of the field notes of an amateur ornithologist (Folke Lindgren)

3 For further information on behavioural ecology, see Paul Martin and Patrick Bateson, *Measuring Behaviour: An Introductory Guide* (Cambridge: Cambridge University Press, 2007) and Peter M. Kappeler, *Verhaltensbiologie* (Göttingen: Springer Spektrum, 2020).

> from 1952 to 2009 and data collected by the Siberian Jay Project (1989–2011). Folke Lindgren (FL) visited the territories repeatedly every year starting in February, to confirm the presence of jays. If birds were present, he regularly visited each group throughout the breeding season (March–May) to catch and colour-ring birds, locate nests and ring nestlings.[4] In autumn, FL revisited the territories to assess how many fledglings remained in the natal group. We confirmed the data on breeding success with help of the records of the Bird Ringing Centre at the Museum of Natural History (Stockholm).[5]

They still do much of what they describe here in relation to Lindgren. The biologists do fieldwork twice a year in spring and autumn to ring all unringed birds and assess how many birds have survived from the previous season. However, they have now developed and formalised their observation methods.

In the 1980s, trained biologists began joining Lindgren. Among them was Professor Dr Jan Ekman, from the Department of Zooekology, now known as the Department of Ecology and Genetics, at Stockholm University in Sweden. Following the encounter between the two, Lindgren's hobby was formalised as the *Siberian Jay Project* at Uppsala University, receiving government funding from 1989. Michael joined the group in 1998. In 2004, now based at the University of Stockholm, he took over the responsibility of fieldwork from Jan and introduced new ways to formally

4
Folke Lindgren, 'Iakttagelser Rörande Lavskrikan *(Perisoreus Infaustus)* Huvudsakligen Dess Häckningsbiologi'. *Fauna och Flora* 70 (1975): 193–232.

5
M. Griesser and S. Lagerberg, 'Long-Term Effects of Forest Management on Territory Occupancy and Breeding Success of an Open-Nesting Boreal Bird Species, the Siberian Jay', *Forest Ecology and Management*, no. 271 (2012): 59.

document data.[6] Michael continued to develop the project in conjunction with master's and PhD students, visiting Arvidsjaur at least twice a year to maintain a complete dataset. However, it was not until 2012 that Michael became the PI of the project when Jan retired. In addition to Michael, my main interlocutor, I met the other biologists who are part of the field study. They include two master's students, Julian Klein and Kate Layton-Matthews, who I met in the spring season of 2015 and two field assistants, Marine Quintin and Camille Toscani who joined in the spring of 2020.

Over time, the number of territories surveyed increased to approximately seventy. By conducting fieldwork, the biologists produced increasing volumes of data and research output on the birds.[7] What appears to be forgotten from the biological perspective is that this knowledge is not only scientifically valid and factual but also implicit and informal, collected by the researchers when working with the birds, and enhancing their scientific knowledge. I aim to make this forgotten knowledge visible again.

As this chapter is called 'Preparing', I focus on the requirements of fieldwork, such as arriving in the field (4.1), preparing for fieldwork (4.2), and learning fieldwork practices (4.3). For this, I draw on my observations while accompanying the biologists during fieldwork. I develop the concept of *situated enskillment* in this chapter to analyse how novices, such as master's and PhD students, learn to do fieldwork at this formal level. The concept of situated enskillment

6
Papers published about the Siberian jays since then are, for example: Michael Griesser and Jan Ekman, 'Nepotistic Alarm Calling in the Siberian Jay, *Perisoreus Infaustus*', *Animal Behaviour* 67, no. 5 (2004): 933–39, https://doi.org/10.1016/j.anbehav.2003.09.005; Michael Griesser, 'Referential Calls Signal Predator Behavior in a Group-Living Bird Species', *Current Biology* 18, no. 1 (2008): 69–73, https://doi.org/10.1016/j.cub.2007.11.069; Michael Griesser, Peter Halvarsson, Szymon M. Drobniak, and Carles Vilà, 'Fine-Scale Kin Recognition in the Absence of Social Familiarity in the Siberian Jay, a Monogamous Bird Species.' *Molecular Ecology* 24, no. 22 (2015): 5726–38; Filipa Cunha and Michael Griesser, 'Who Do You Trust? Wild Birds Use Social Knowledge to Avoid Being Deceived', *Science Advances* 7, no. 22 (2021): 1–6, https://doi.org/10.1126/sciadv.aba2862.

7
Griesser and Lagerberg, 'Long-Term Effects of Forest Management on Territory Occupancy and Breeding Success of an Open-Nesting Boreal Bird Species, the Siberian Jay', 59.

relates to Haraway's situated knowledges, which is knowledge that is relational and dependent on its context. Enskillment refers to the bodily and sensory skills that must be learnt. I refer to Barad in my analysis of how research objects and subjects intra-actively co-constitute one another.

My analysis includes everyday life in the field, with a focus on how novices learn to collect data. This is analysed with reference to wayfinding practices that the biologists must develop to navigate the field and find the birds. To discuss the concepts and parameters of wayfinding practices (4.3.1) and a specific kind of cognitive knowledge, I refer to Kirill Istomin and Mark Dwyer, and Tim Ingold's research on the relationship between the body and the environment.

In the second part, I discuss the registration of birds (4.3.2) as one specific situated enskillment based on my observations of how Michael taught Marine and Camille to catch and register birds. To describe the kinds of practices and knowledge necessary for this step, I refer to Barad's concepts of touching and boundary-making. I question these as situations in which nature–culture dichotomies are transformed into naturecultures. Last, I turn to Thyssen and Grosvenor's (2019) concepts of sensory and bodily learning to discuss how learning environments are shaped by sensory enculturation. There I address the actual data collection, which follows a situated enskillment, in Chapter 5 of this book.

1

2

3

4

5

6

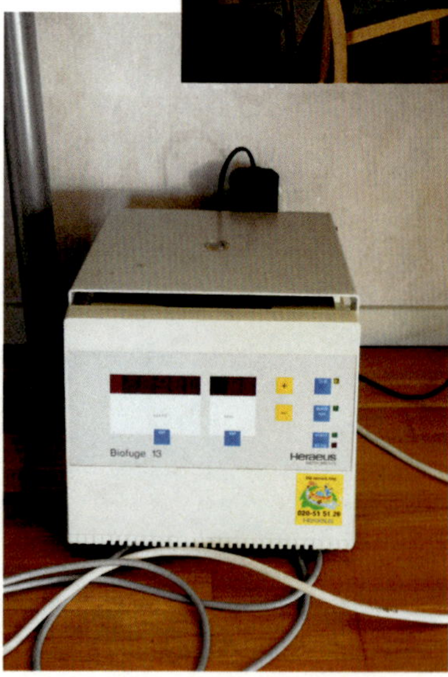

7

8

9

Figure 1:
Arrival of Michael and me at Arvidsjaur Airport in the late afternoon. Arvidsjaur, 2015.

Figure 2:
The field house. Arvidsjaur, 2015.

Figure 3:
Having just entered the field house, looking to the right: A hand-carved fox and cat, fridge with hand-made figures, breadmaking machine, and kettle in the background. Arvidsjaur, 2015.

Figure 4:
Living room. Arvidsjaur, 2020.

Figure 5:
Photograph of a drawer in the field house with tools and media used for fieldwork. Arvidsjaur, 2015.

Figure 6:
The 'office-corner' with tools, media, and traces of everyday life. Arvidsjaur, 2020.

Figure 7:
Unused, old centrifuge. Now, blood samples are stored in a tin can (Figure 38 e) in a freezer and transported back to the universities for analysis once a year. Arvidsjaur, 2015.

Figure 8:
Kitchen from the inside on a day off. Arvidsjaur, 2015.

Figure 9:
Michael indicating an imaginary line where the study site starts while driving to the study area *Managed* on a snow-covered road. Arvidsjaur, 2020.

4.1. Arrival

Once Michael and I arrive in Arvidsjaur (Figure 1), we find ourselves in a peaceful, snowy landscape, where buildings are scarce and the winter sun makes all surfaces glisten. When we leave the airport building in Arvidsjaur, a couple around my age are waiting for us on the bonnet of the field car. They introduce themselves as Julian and Kate and mention that they arrived from Zurich a few days earlier. Kate is a master's student in the Department of Ecology at the University of Zurich and Julian is a former master's student of Michael's in the Department of Anthropology at the University of Zurich, where Michael is employed as a Swiss National Fund research professor at this point. This scene will repeat similarly the following season when I meet Marine and Camille, volunteer field assistants who had previously studied biology.

Before continuing our drive to the field house, we stop at the local supermarket to buy some groceries. Besides the groceries that will sustain us during the next few days of fieldwork, our shopping carts are filled with several packs of fatty sausage, wrapped in an orange-red plastic and covered with pig fat and skin. Eventually, we simply refer to this as 'fat' or 'feeder', as it serves as food for the birds and helps to attract them. More precisely, however, the term 'feeder' refers to a stick onto which two pieces of fat are attached.

Once we finish our shopping we continue driving through the snow-covered landscape of Arvidsjaur, our boot filled with the (still) frozen fat and sausage. On our way towards the house that serves as a field camp, it becomes clear that Michael has now entered the field. During the journey, he starts discussing the conditions of the field with Julian, who had been in the study area the previous year, and what they had been able to prepare thus far. He frequently peeks outside and studies the sky, trees, and forests as we pass. He seems to be looking for birds or changes in the landscape attributable to forest management, preparing himself for what he will encounter the next day when we start our data collection from the jays.

While Michael first had to arrive in Arvidsjaur to enter the field, I had entered the field a few months earlier when I visited the biologists' office, where they were still evaluating data from the previous seasons. I had casual conversations with Michael about the use of drawings, colour coding, and maps, and how evolutionary biology constructs knowledge based on translation processes supported by visual aids that transport the research object to research offices, similar to how Latour described the process in 'Circulating reference: Sampling the soil in the Amazon Forest'.[8] Subsequent office visits involved planning the fieldwork: clarifying the biologists' research questions and objectives, outlining data-collection methods, determining my preparations, and defining our collaboration. Now, the theoretical concepts I had explored in the university offices were about to be translated into practice.

The Field House

After a 20-minute journey, Julian makes a left turn, and we enter a driveway. The beams of the car light up a beautiful wooden Swedish house, like those I have seen online and in travel magazines. We have arrived at the field station (Figure 2). The field station – recently named *Luondu,* the Sámi word for nature – has long vertical wooden panelling. Its south-facing wall is painted white, blue and yellow, which makes it appear warm and inviting. It is located on a property close to one of the main roads. The road was quiet on our arrival, although occasionally large trucks drive past, their bright lights shining to avoid running over any reindeer crossing the road.

8
Latour, 'Circulating Reference: Sampling the Soil in the Amazon Forest'.

To the left of the house is a smaller cottage, named after the Sámi term for growing lichens, 'goavsak biesse'. It is equipped with a stove, useful during the initial days of fieldwork when the main house still needs to heat up. A large building stands behind the house and includes the garage, a storage room for field equipment, and a room for firewood, as well as another small wooden cottage. The lighter colour of the wooden beams that cover the house gives the impression that it was built more recently. Inside, one sees that it functions as a sauna. Next to the large building there is an adjacent outhouse. The neighbours' houses can be spotted faintly through the trees; behind the houses is a large empty space covered in snow, making it difficult to discern whether I am looking at a frozen lake or a snow-covered field.

We enter the main house through a porch with windows; this is where we would eventually store our field equipment. On entering the house, I find myself in an open corridor that is separated from the kitchen with a low wall and a wooden balustrade on top of it (Figure 3). There is a bathroom to the left of the corridor: a long, narrow space with a toilet, sink, and shower. The bathroom and corridor were refurbished with a wooden cover between my two field studies in 2015 and 2020. Michael seems to be constantly improving the house together with friends and assistants.

Next to the bathroom is a small nook with a washing machine, and at the end of the corridor another door leads to the first of three bedrooms. The living room was converted into an office in 2024, but in 2020 it was still one large space with futons and fieldwork equipment (Figure 4). In one cupboard, I find several smartphones and tools that Michael will use during the study (Figure 5). Walking through the room, past the big window on the right that faces the main street, I find the door to my room. To the right, there is another nook that functions as an office (Figure 6), with a desk and chair, cables, camera chargers, gear for the fieldwork, an unused centrifuge (Figure 7), and a shelf of books on behavioural biology. The third bedroom is to the left.

The kitchen is the main meeting point and workspace in the house (Figure 8). This is usually where the field-day planning and discussions on previous excursions take place. A small router with poor Internet connection is installed here; in 2020, it was replaced with a smartphone with a hotspot that requires one to be close to get a signal. Access to the Internet and digital notebooks does not appear to be a top priority here. The only reason we use the Internet router is to occasionally send emails or call our loved ones. Michael also video calls his colleagues at the house when he is not able to join them in Arvidsjaur.

A table with some chairs and a bench stands in a corner. A large fridge appears to be placed almost in the centre of the room, with a floor hatch in front of it that leads into the basement, where we store our groceries. There is also a freezer, and we fill several of its drawers with the frozen fat and sausage. The field house is a homely place, and everyone appears to be comfortable from the moment they arrive, which is important for productive collaboration. Michael tells me that a good group dynamic and team are essential to collecting valuable data. The group requires pragmatic people: skilled fieldworkers and those who do the 'care' work. 'Care' in this context refers to those that take care of the cooking and have the capacity to create a homely atmosphere, while others might be more technically focused, preparing tools and organising fieldwork. Researchers' children also sometimes come to the field house and need looking after.

Establishing Relations

Michael tells us that he renovated the field house with the help of friends with carpentry skills who were paid, but in turn could live in the house for free while working on it. There seems to be a correlation between the people who come to Arvidsjaur to do fieldwork and having these types of skills. Julian is not only a biologist but also a trained plumber; he helped fix the plumbing in the field house. Filip, a friend of another PhD student, helped renovate the house and stayed there during the off-season. Currently he is occupied with building an additional cabin in the back of the land so that Michael can host more biologists and field assistants in the coming seasons, or the accommodation could be used for those that bring families along and require a more private space. Michael also fixed, installed, and built several features in the house, such as the new wooden porch at the entrance and the sauna.

Thus, the field house is not just any Swedish house. Michael owns it and comes here often, and its features tell stories about the people who have stayed here and helped to build it. The house serves as an archive of stories about the *Siberian Jay Project*.[9] The features of the field station, *Luondu*, are not simply the result of functional decisions; further details of the house reveal a social network that goes beyond biologists doing fieldwork. The way Michael uses the house renders it not just his house, but the house of a (scientific) community centred around a common interest in fieldwork. This may be from both a biological and artistic perspective; photographers, filmmakers, and musicians have also joined Michael here in the past. Some of them were not only there as artists but also inscribed themselves into the house through their carpentry skills.

9
When Jan Ekman took over from Folke Lindgren and secured funding from the Swedish Research Council between 1989 and 2011, the project was named the 'Siberian Jay Project'. While the research no longer receives the same funding, the name is still used informally when discussing the project.

During our stay with Michael over the course of several weeks, he shares stories of people who have left their mark, for example, a friend who is a photographer. One day, when I visited my neighbour in Zurich, I found an old black-and-white photograph of Michael on their kitchen table, taken in Lappugglan in Arvidsjaur in 2003. It turns out that my neighbour is the partner of the photographer who had taken the photograph of Michael in the early 2000s. In this way, new connections are continuously being made, and new stories can be told, densifying the network of people around Michael's field house and the research project beyond scientific results and publications.

These are not the only anecdotes about Arvidsjaur; there are others about the traces people have left in and around the field house. Moreover, the study sites are shaped by the people engaged in the study. Each bird territory receives a name, and these names are often related to stories about events that occurred during fieldwork.[10] Thus, they serve as an oral history of the research project and as mnemonic techniques to aid memory of a specific territory's name. The stories are passed on by each new generation of biologists and continue to circulate. They reveal insights into the environment in which the territories are situated and tell stories about events, shaping the identity and culture of the team.

Anecdotes

Lord, Troll, Fat Moose, Fat Jana, Take 5, Impossible, Akkavare, and *Angel* are examples of territory names that circulate among the group but are not usually published in scientific journals. These names are inspired by a variety of

10
Anecdotes that give insight into the origin of selected territory names are presented at the end of the book. I collected these partly during fieldwork and partly during interviews with Michael. As the territories are constantly developing, with some disappearing and new ones appearing every year, I included only the main ones I refer to in this work.

sources: natural phenomena, such as animals that have been spotted (e.g. *Fat Moose*); characteristics or names of people involved in fieldwork (*Nadine, Kara, Fat Jana, Basilika, Impossible*); films – particularly the black-and-white movies of Fritz Lang, on which Jan was an expert (*Metro* and *Mabuse*); books being read during fieldwork (*Lord* and *Baggins*); memorable incidents while working with the birds (*Angel, Take 5, Blot*); and notable landmarks or traditional place names (*Akkavare, Mader, Måskomyran*). They also reflect the changing culture of the research team. In the early days of the Jay project when Jan was the main researcher, the names he chose for territories were strictly Swedish. More recently, the territory names show a shift to names that are not exclusively Swedish, reflecting the more diverse cultural backgrounds of the participating researchers. Thus, the maps with the territory names that also appear in Excel sheets and in field notebooks are not only valid as scientific data but also represent a dense network of narratives about the research conditions and other unique and exceptional aspects, thus working as mnemonic techniques. As territory names, they remain in the reference system established by the biologists and continue circulating throughout the study.

The territory names work as immutable mobiles and create consistency. Their narrative nature means they are valuable data for an STS and design analysis. They do not become data, but as part of the reference system, they add meaning to individual data points. Data become data only in reference to other data; thus, they always gain their meaning in relation to these immutable mobiles, creating entanglements between research objects, data points, and results in a way that the biologists can make sense of. However, this is a side effect. On the one hand, the biologists aim to categorise nature with objective tools and methods, but on the other, they must use these methods to master the field, maintain an overview, and organise the study.

Biologists working with this group over time can usually recall the events to which the territories owe their names. Experienced biologists remember specific events of the study. Michael, in particular, can recall almost every anecdote or origin of the territories' names. The biologists use these narrative units as a very effective mnemonic technique to support their research, as it helps them remember and discuss the specific territories more easily. However, they never seem to reflect on its impact on their research practices.

The territory names and side-note entries that the biologists make in their field notebooks are one of the few instances where the emotional, sensorial, or social conditions of this biological study become visible beyond ephemeral moments during fieldwork. In most other cases, the biologists must 'disenchant'[11] the birds and the boreal forests for the sake of scientific integrity, robust data collection, and development of valid models of their encounters. However, with this name 'trick', the biologists themselves briefly undermine the scientific norms of objectivity, neutrality, and reproducibility with which they hide themselves in the scientific process. This time, though, the trick is one that situates them within the network of their own stories rather than outside it.

When it comes to publishing papers, these colloquial names of the territories are replaced by more formal ones, such as number codes that filter out these aspects and eliminate the names from the scientific discourse, along with the associated stories (Chapter 6). Lastly, from an STS perspective, these names reveal insights into the conditions of knowledge production and the overlap between the two ontologies, nature and culture, which are usually conceptually divided in theoretical discourse.

11
Kohn, *How Forests Think: Toward an Anthropology Beyond the Human*, 90.

Shifting Meanings of Spaces

Different spaces have different meanings to the biologists and to me. For the biologists, the field house is, in the first place, their accommodation, a storeroom for their equipment, an office, and a meeting room; to me, it is part of the field, and so are the car, supermarket, village, boreal forests, study site, university offices, and online calls.

One location can serve several functions depending on the person who enters it. To the tourist, the forests are a place of leisure and vacation; for local residents, it is home and part of their everyday life; to the Sámi people, the forests are part of their agricultural economy, like reindeer herding; to landowners, the forests must be 'managed', and for the biologists conducting research, it is their work environment and study site: 'the field'. Michael always points out when we enter 'our study site', crossing an invisible border between the town of Arvidsjaur and the village of Auktsjaur where the field house is located (Figure 9). When speaking to one another, the biologists do not speak of 'the forest'; instead, they call it 'territory', 'field', 'study site', or they use the unique names assigned to the three main areas, *Managed, Fat Road,* and *Reivo.*

Occasionally, in places such as the supermarket, the different groups of people come together with the same goal of buying groceries. One encounters locals doing their grocery shopping after work, alongside the biologists and me (wearing field clothes with big boots, gaiters, and many layers of insulating and waterproof clothes), and seasonal workers from the car-racing track in Arvidsjaur, situated near one of the study areas, disturbing the winter silence with the engine noise of their racing cars.

The boundaries between the field, fieldwork, and everyday life in Arvidsjaur differ for the biologists compared with me. I collect data on aspects concerning both work and everyday life because fieldwork is only possible through the combination of the two. From an STS perspective, both serve as data collection. To the biologists, data collection occurs only during fieldwork, and even then, only a small number of observable aspects become data when working with the Siberian jays. To them, other practices are an inevitable part of preparation; they are not data collection.

Accordingly, depending on the expectations and the purpose with which one visits Arvidsjaur, the concept of what this place is shifts and, with it, how realities are constructed around it. In our case, Arvidsjaur can be viewed from two research perspectives, biological and ethnographic. Therefore, it must also be understood as a space of twofold knowledge production, as I elucidate in the following section.

10

11

12

13

14

15

Guorbåive
Storberget
Guoratjliden
NJÄLLÅVARE
Måskape
Lord
Laxfors
Baggins
Fyra
Granmyran
Metro
MI5
Lägerplats
Reivosjö
Mabuse
Täppan

16

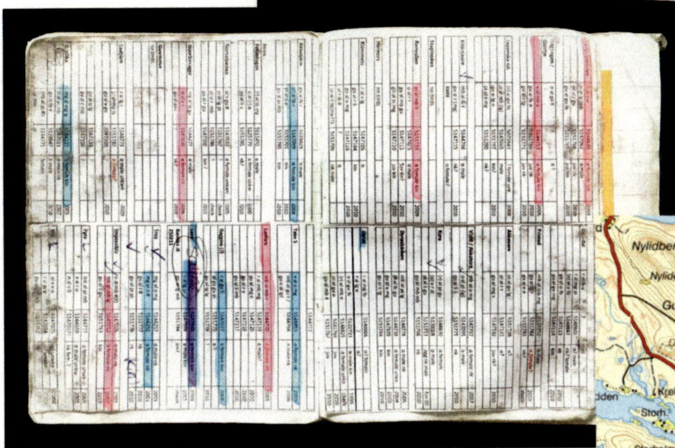

17

Nylidberget
V Såkevare
Tellekflygget
Kvarnberget
Nyliden
Ö Såkevare
Stor-Bergmyran
Guortesliden
Dyrasflygget
Domänreservåt
Såkevaretorpet
Dyras
392
Långmyran
Måskomyran
Storholms-berget
366
Glotte
Akkavareliden
534
Björkliden
Akkavare
Flarkmyran
45
Dammberget
Grandal
Abborrtj.
Bergsviken
Renudden
Fristad
Gravatjärnen
Stentrask-berget
367
Arvidsjaursjön
Renvallen
Klöcksta
364

18

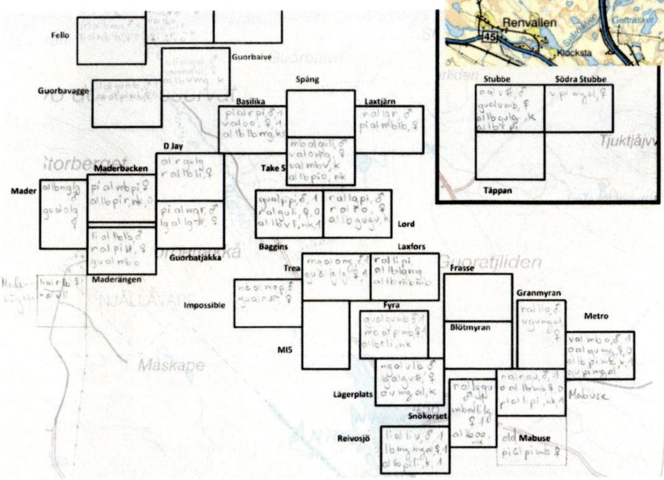
Fello
Guorbaive
Spång
Guorbavagge
Basilika
Laxtjärn
Stubbe
Södra Stubbe
Täppan
D Jay
Maderbacken
Take S
Mader
Lord
Baggins
Laxfors
Guorbatjäkka
Maderängen
Trea
Frasse
Impossible
Fyra
Granmyran
Blötmyran
Metro
MI5
Lägerplats
Snökorset
Mabuse
Reivosjö
Måskape
Guoratjliden

19

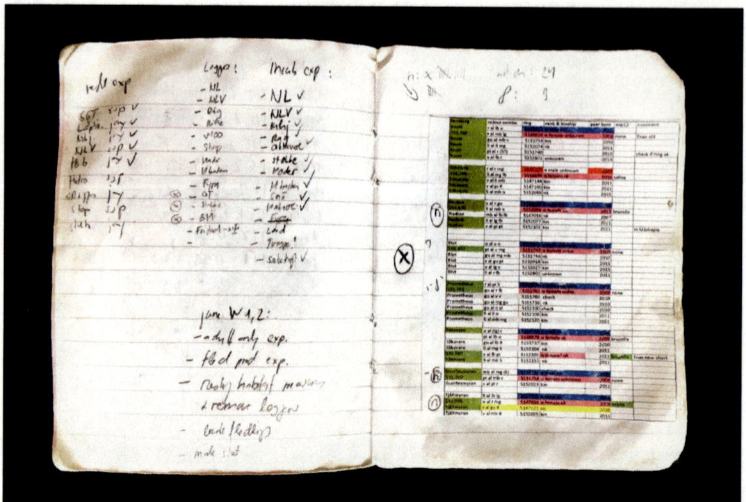

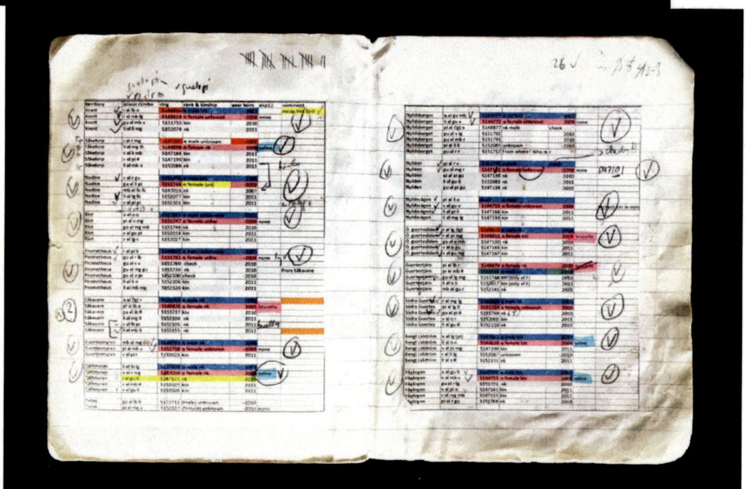

20 a, b, c

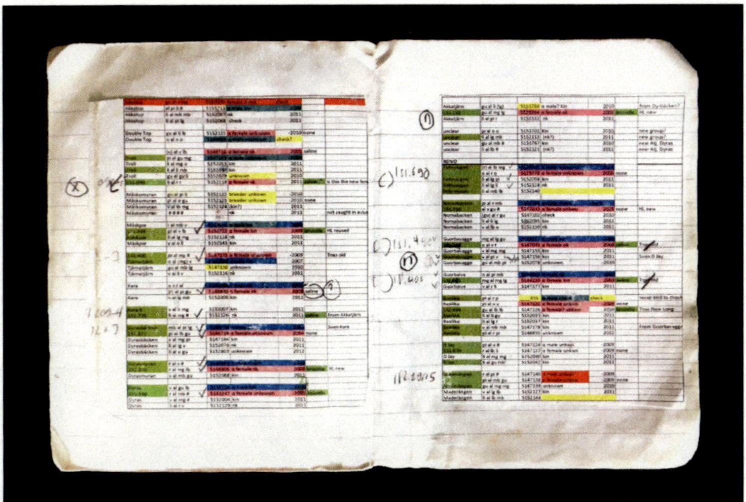

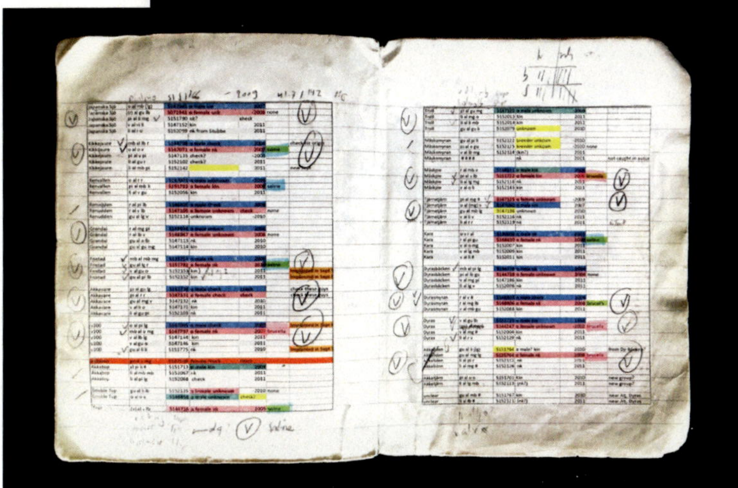

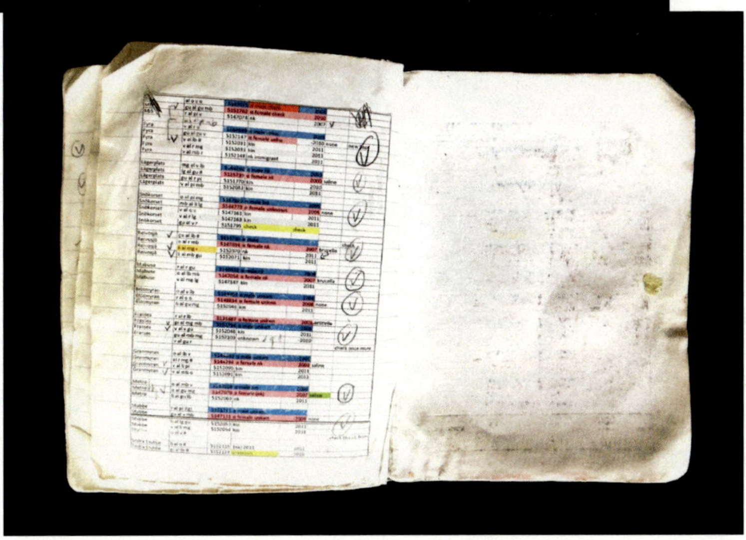

20 d, e, f

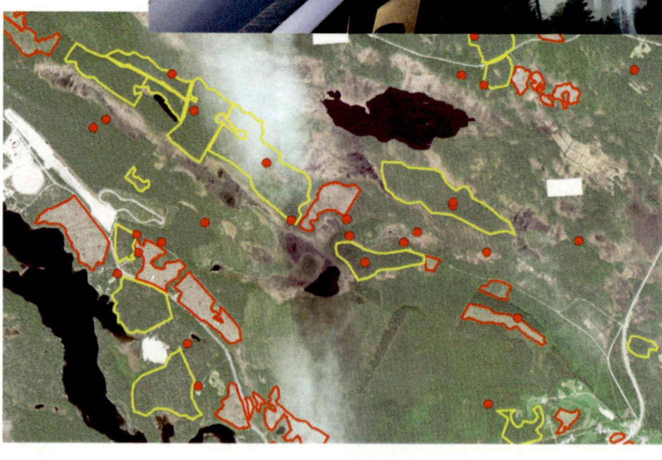

21 a, b

Figure 10:
Michael preparing the gear for the field day, fixing the binding of one pair of skis in the morning before fieldwork. Arvidsjaur, 2015.

Figure 11:
Fire stove that runs the heating system in the kitchen next to the cooker. Arvidsjaur, 2015.

Figure 12:
Firewood to run the stove and birch bark to start the fire, just brought in from the wood storage in the garage at the field house. Arvidsjaur, 2015.

Figure 13:
Prepared feeders for the Siberian jays, incised with a knife to make it easier for the birds to feed, and with two holes and wire threaded through them to attach to trees. Arvidsjaur, 2020.

Figure 14:
Kitchen stove with a coffee machine on the fire. Arvidsjaur, 2020.

Figure 15:
Marine, Camille, and Michael (l.–r.) standing in the kitchen in front of a cork board with the three territory maps, discussing and planning fieldwork for the day. Arvidsjaur, 2020.

Figure 16:
Double page in the front of Michael's notebook from 2010 with territory maps, markings, and notes of the territories that have already been visited. Zurich, 2018.

Figure 17:
Back of a student's field notebook from the 2011 season with a printed field list glued into the notebook. Zurich, 2018.

Figure 18:
Map of study area *Managed* as it is used for the territory maps. Zurich, 2018.

Figure 19:
Map of the study area *Reivo,* with boxes representing the individual territories and the relevant names near the boxes. Handwritten code with pencil, referring to the IDs of the individual Siberian jays that live in the respective territories and have been seen that season. Arvidsjaur, 2015.

Figure 20 a–f:
Field lists of all Siberian jays, glued into Michael's field notebook of 2012 with handwritten code to indicate the territories visited and the tasks fulfilled.

Figure 21 a, b:
Getting ready for fieldwork. Michael and Julian fixing the skis onto the roof of the field car. Barbara, who visited her partner Julian for a few days, and Kate, are waiting. Arvidsjaur 2015.

Figure 22:
Map of forest management in the study area *Managed.* Markings denote different management types. ArcGIS. Zurich, 2018.

4.2. Preparation

Days start early in the field camp, with our alarms going off at around 6:20 a.m. Before someone starts the fire in the mornings, the rooms are chilly. I slip into my insulation clothes, which are on the floor next to my bed so that I can warm them up before completely leaving the bed. While my roommate Kate and I struggle to rise this early, Michael and Julian are usually already in the kitchen by now.

When I enter the kitchen, Michael, who usually gets up first, is busy with the bread we baked overnight in a breadmaker. In Arvidsjaur, a breadmaker receives more appreciation than a fast Internet connection or smartphone. However, this has changed since our fieldwork in 2021. These days, Michael tells me that people are 'glued to their phones' in the mornings, since WiFi access has improved in the field station. Once Michael has removed the bread, he starts preparing whatever else is necessary for the day (Figure 10). Most often, this is the equipment. Michael usually decides what must be done during fieldwork and thus has the best knowledge of the equipment required. I quickly check the thermometer that is installed next to the kitchen window to check how cold it is today. It indicates –17°C, a little milder than usual in the mornings.

Julian sits on a wooden chair and starts the fire (Figures 11, 12). Soon he will move over to the stove and make porridge for everyone. We all appreciate the hot food in the morning because it warms us and prepares us for the field day out in the cold. In one corner of the kitchen, a pile of fat is defrosting (Figure 13). Someone will cut it into smaller pieces, make two holes in the middle and thread a wire through it, so that the biologists can easily attach it to the trees in the territories. In this sense, the fat is part of the experimental system. This task, which no one is particularly fond of, is rotated among us; today, it is my turn.

After a few days of acclimatising to the field and the morning tasks, Kate makes coffee (Figure 14), while I set the table for breakfast, prepare the thermos flasks, and pack our lunch with fresh bread. I often do this while the biologists finalise the planning for the fieldwork. Their aim is to visit the roughly 70 registered bird territories in the three study areas at least twice during this season. During the first round, they mostly check on the birds and document their first observations. Later, they will make video recordings and, if it is planned for the season, conduct further experiments. They may also have to return to territories to ring new birds and register them; or, if the birds did not show up the previous time, they may go back to attempt to find them again.

The group decides who collects data in which bird territory that day by evaluating the most efficient routes and what must be done. For this task, they sit together at the kitchen table, studying the maps and discussing potential problems as well as the likelihood that birds will appear. The group also take their individual skills and experience with fieldwork in Arvidsjaur into account during the division of labour. In this way, they have a realistic plan with tasks that everyone can manage based on their abilities. It is easy to split the territories among the biologists on the first day when no data have been collected yet. In 2020, the planning was done in a similar manner: Michael, Marine, and Camille would discuss the plan for the day, often looking at a map that Michael had put up in the kitchen after a few days of fieldwork (Figure 15). For these planning discussions and considerations, the three biologists stand around the maps, while I sit on the kitchen table behind them and observe them. Michael points towards the territories, while Camille and Marine are standing next to each other with their field notebooks in their hands: listening, discussing, and taking notes. Camille and Marine occasionally confirm what he says by repeating the territory names, nodding their heads, or simply noting the names in their field notebooks.

Before setting off for fieldwork, the biologists collect, check, and pack their gear. Some days they take mist nets and a tacklebox to ring the birds and collect data on their size and weight. This usually happens only after a few field days, as at the beginning of a field season they first attempt to obtain an overview and focus on the presence of the birds. On these days, they carry their standard equipment, which consists of the field notebook containing the lists of the birds and maps of the territories (Figures 16–19), a pencil, binoculars, bait, and an old phone – often a Nokia smartphone (because their batteries last the longest) with a Swedish SIM card. On other days, this equipment is supplemented with a video camera, or, when it comes to conducting experiments, the biologists carry additional tools with them. Depending on the plan for the field day, the gear, material, and tools change, as do the morning preparations. As I need only my field notebook, pencil, camera, and smartphone, I help prepare or carry the biologists' equipment.

On the first evening in the field camp, Michael provided me with a field notebook, like the ones the biologists use, and a pair of binoculars. With Julian's guidance, I glued the three maps of the study areas at the front of the small A6 notebook. At the back, I glued in an Excel list (Figure 20) with all the birds that had been registered for the study which the biologists suspect are currently living in the territories. Gluing these maps and charts into the notebooks is something all the biologists do before they enter the field. Thus, they make sure that they do not lose their most important documents for fieldwork, as the combination of the maps and territory lists helps them know what to expect and which birds should be present. During fieldwork, I make less use of the notebook and check the maps only occasionally to see which territory we are in; however, without guiding myself and without a GPS, the maps alone are not much help, and I often have to ask my interlocutors which territory we are in. Most often, I stick to the notebook in which I collect my ethnographic data.

Once we have finished breakfast and prepared our equipment, we store the backpacks containing the – now defrosted – fat, poles, and other tools in the field car. Our skis are attached to the roof, and the biologists and I get in, ready to set off (Figure 21). A quick glance at the clock through the driver's headrest reveals that it is no later than 7:30 a.m. and no warmer than about –19°C. The driver then obscures my vision of the clock.

Michael does not start driving without checking one last time that everyone has packed their notebooks and pencils, the biologists' most essential tools for fieldwork. However, the thermos flask is equally important and also forms part of the checklist. Once it is confirmed that nothing important has been left behind, Michael starts driving towards one of the three study areas that has been chosen for the day. On other days Julian drives. The drive takes between 20 and 40 minutes, depending on the site. Time is precious during fieldwork, as the return flights have been booked and the biologists usually have a tight schedule in which to complete many tasks. This urgency is particularly clear in Michael, who does everything quickly – including driving.

Three Study Areas

The field consists of three main geographical zones, study areas, or study sites, as the biologists call them, represented by the three maps in my notebook: *Managed, Reivo,* and *Fat Road.* In turn, the study areas contain several 'bird territories'. Each territory is inhabited by one group or 'family' of birds. The borders of the study areas are defined either by natural markers, such as hills and rivers, or by roads, such as in the territory *Managed,* a managed forest – from which its name is derived – that is encircled by two roads. However, the borders change slightly depending on the evolution of the bird territories, as the birds apparently do not remain in human infrastructure. Therefore, there is an additional territory across the road from *Managed* to which the birds have migrated – something that occasionally occurs.

Reivo is a hilly, protected forest, situated alongside a valley. It is divided into two parts by a ridge that we must cross when collecting data. While the territories on the near side between where the car is usually parked and the ridge are within 30 minutes' skiing distance, the others, on the far side of the ridge, only start after about 30 minutes of skiing. Therefore, most territories in *Reivo* require a greater level of physical fitness and skiing ability. At the same time, it is also the most beautiful forest of the three, which, at least for me, is motivation and makes up for the effort required.

The *Fat Road* study area is located along a forest road. Working here may also involve a long skiing route, but the territories are easily accessible because they are all located alongside the road, which makes the skiing and wayfinding easier than when navigating through the forest. When the biologists first searched for birds in that region, they noticed that they had created a 'fat road' by putting up a piece of fat to attract the birds every kilometre, and thus they decided to call it *Fat Road*.

Each study area contains bird territories in which bird 'families' are located. Family – one of many anthropomorphic concepts that have been applied to these birds – means that the breeding couple, a male and female bird, usually has offspring every year. The offspring are called 'nestlings' and, once they have left the nest, 'kin juveniles'. As 'kin juveniles', the more dominant birds chase their submissive siblings away and retain the territory themselves. The remaining birds then stay together in a family network of three to four kin birds. The juveniles that have been 'kicked out' of their birth territories usually join other bird families. Thus, some birds live as adopted 'non-kin' birds or unrelated 'immigrants' with these family groups, forming a 'patchwork family' of sorts. After a year or two, the 'kin offspring' leave their parents' territory to find a mate. Subsequently, they either form a new territory together, or, as the breeding couple usually remain together for life, they join an existing territory in case one of them dies.

As jays are territorial, the breeding couple do not change territory, nor does the rest of the group, until they become breeders themselves or are chased away. The birds mostly move within their territory, which spans approximately 800 m. This is where they hide their food in winter and build their nests in spring. More recently, the territories have been increasingly affected by deforestation and some bird territories have disappeared entirely.[12] I was particularly struck by the impact of this in the 2020 study, where, in comparison to my study five years prior, much of the forest had been felled and some territories we wanted to visit had disappeared (Figure 22). However, before we enter the field and can collect data, the biologists and I must receive situated enskillment.

12
Kate Layton-Matthews, Arpat Ozgul, and Michael Griesser, 'The Interacting Effects of Forestry and Climate Change on the Demography of a Group-Living Bird Population', *Oecologia* 186, no. 4 (2018): 907–18, https://doi.org/10.1007/s00442-018-4100-z.

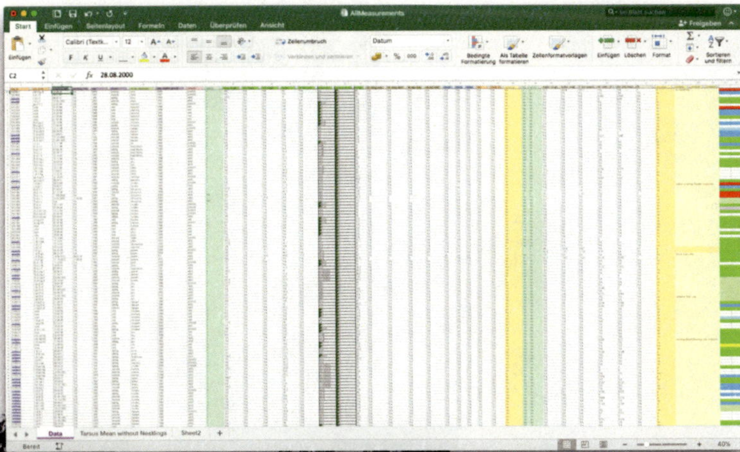
AllMeasurements
Start
Einfügen
Seitenlayout
Formeln
Daten
Überprüfen
Ansicht
28.08.2000
Data
Tarsus Mean without Nestlings
Sheet2

23

24 a, b

24 c, d, e

25 a, b, c

26

27

28 a, b, c, d

29 a, b, c, d, e

31

32 a, b

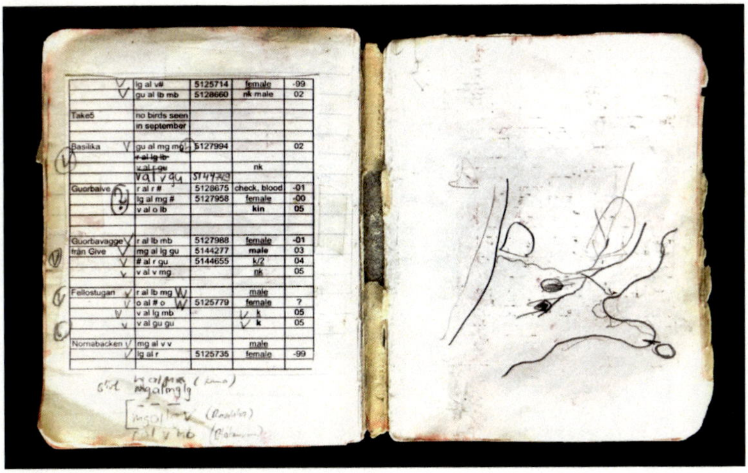
lg al v# 5125714 female -99
gu al lb mb 5128660 nk male 02
Take5 no birds seen in september
Basilika gu al mg mg 5127994 02
r al lg lb
v al r gu nk
Guorbalve r al r # 5128675 check, blood -01
lg al mg # 5127958 female -00
v al o lb kin 05
Guorbavagge r al lb mb 5127988 female -01
från Give mg al lg gu 5144277 male 03
al r gu 5144655 k/2 04
v al v mg nk 05
Fellostugan r al lb mg male
o al # o 5125779 female ?
v al lg mb k 05
v al gu gu k 05
Normabacken mg al v v male
lg al r 5125735 female -99

33

34 a, b, c

35

36 a

36 b

36 c

36 d

36 e

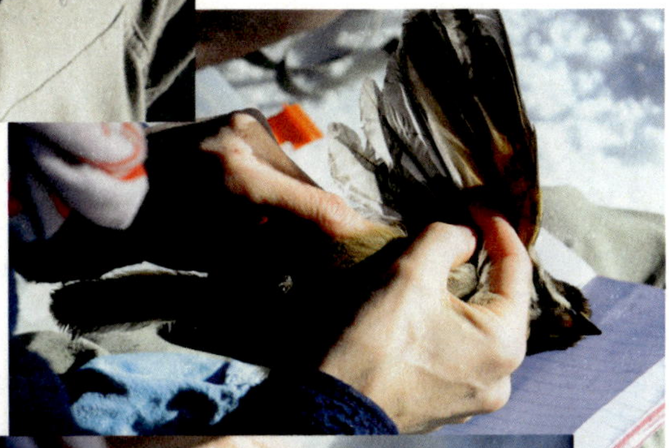

36 f

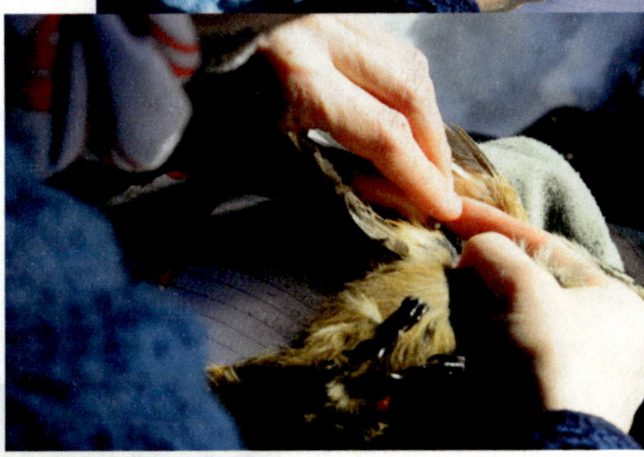

36 g

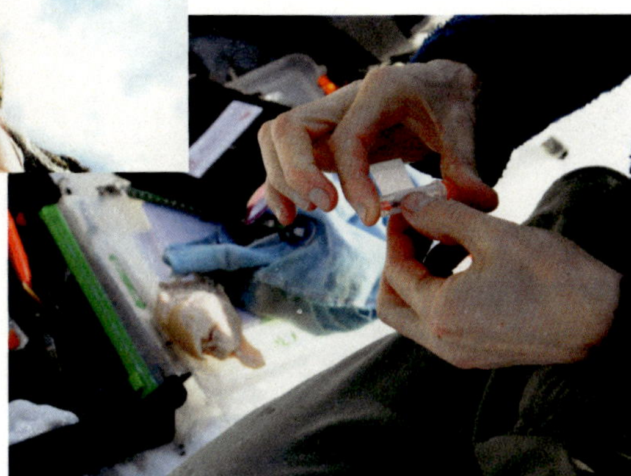

36 h

36 i

36 j

36 k

36 l

36 m

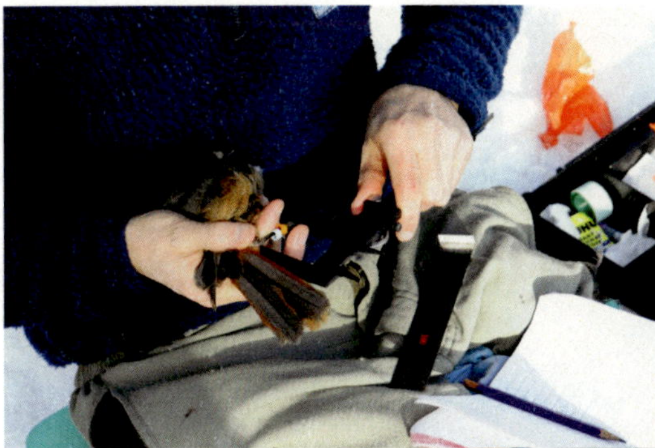

36 n

36 o

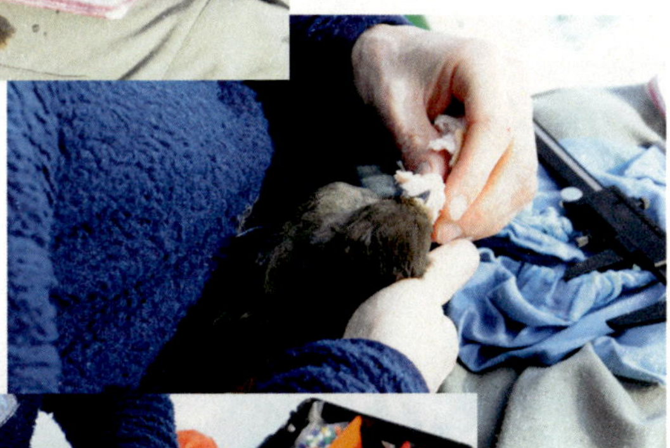

36 p

36 q

37

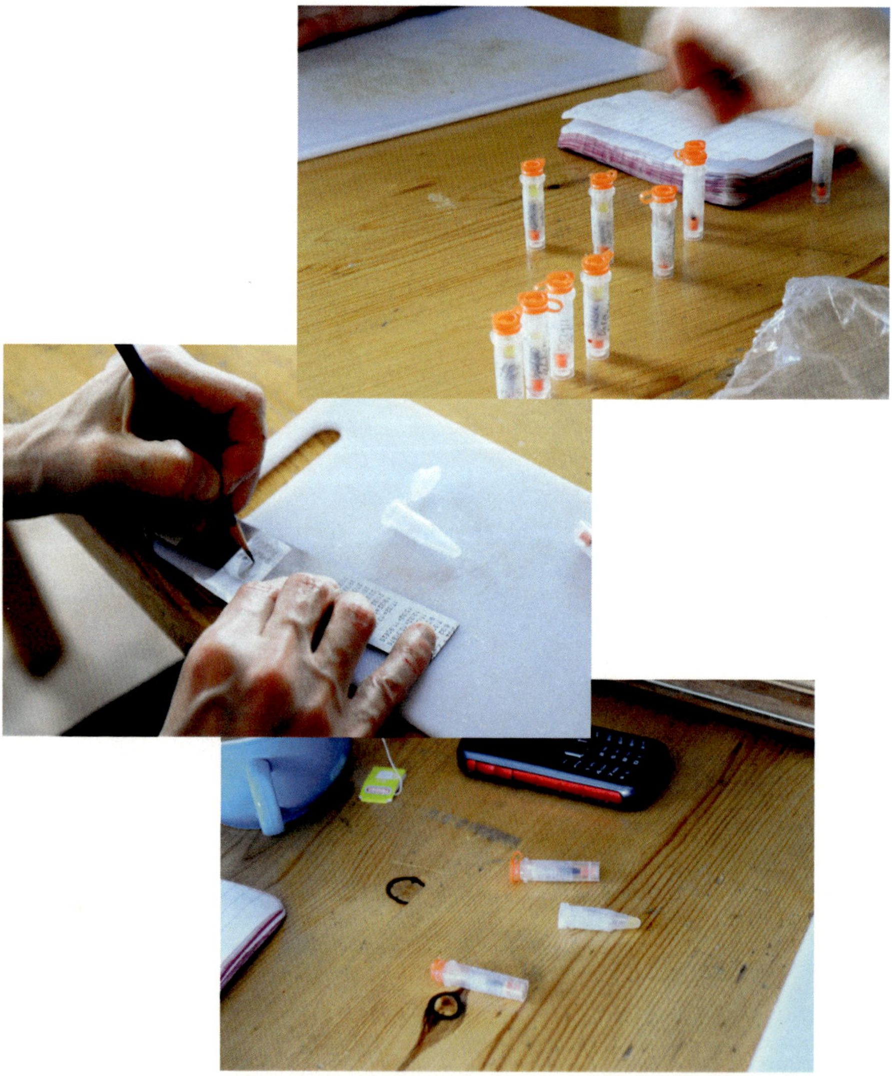

38 a, b, c

38 d, e

Figure 23:
Screenshot of an Excel sheet of the main database containing all measurements (obscured to avoid theft of data). Zurich, 2017.

Figure 24 a–e:
Tracks made by Kate and me. Arvidsjaur, 2015.

Figure 25 a–c:
Sequence of Michael on cross country skis, putting up a feeder on a tree. Arvidsjaur, 2015.

Figure 26:
Feeder impaled on a branch with lichen. Arvidsjaur, 2015.

Figure 27:
Camille holding the GPS set up to go to territory *Trollhunter* 2019, usually called *Troll*. Arvidsjaur, 2020.

Figure 28 a–d:
Sequence of Marine during her first video recording of a behavioural observation, feeder attached to a tree in the back, currently with one bird around. Arvidsjaur, 2020.

Figure 29 a–e:
Sequence of Camille crossing an open field, looking for Siberian jays in a territory affected by deforestation. Arvidsjaur, 2020.

Figure 30:
At the end of a field day in *Reivo,* I try to capture the atmosphere from where the car is parked. An empty house to the left and a farm in the background. Arvidsjaur, 2015.

Figure 31:
Video camera '1' covered with snowflakes and installed in *Maderängen,* recording the interactions of the Siberian jays. (A photograph Michael asked me to take to document the snowflakes.) Arvidsjaur, 2020.

Figure 32 a, b:
(a) Marine, Camille, and Michael planning their field day in *Reivo,* taking notes and discussing the best clusters of territories to visit. Arvidsjaur, 2020. (b) Michael giving Camille more detailed descriptions for her route, showing her the territory location on GPS. Arvidsjaur, 2020.

Figure 33:
Double page of a field notebook, showing the Excel chart with the Siberian jays glued into the field notebook and a hand-drawn map to document a geographical situation during fieldwork. Zurich, 2018.

Figure 34 a–c:
Sequence of a bird flying into a net installed between the trees. Michael in the background. Arvidsjaur, 2015.

Figure 35:
Blue cotton bag holding a bird that has just been caught. Arvidsjaur, 2015.

Figure 36 a–q:
Sequence of taking bodily measurements from a bird. Arvidsjaur, 2015: (a) Michael is about to take out the bird from the blue cotton bag lying on his lap. The field notebook, a pipette for the blood sample, a calliper to take the bodily measurements, and a box with the coloured rings are prepared next to him. (b) He is about to pick a tail feather of the (ringed) bird. (c) Together with this, he checks the tail (and wing) feathers to count the so-called 'hunger bands' that indicate the fitness of the bird. (d) Now Michael notes down the colour ID of the bird into his notebook, (e) holding the bird ready to measure it. (f) Next, he opens the right wing of the bird to take a blood sample by (g) puncturing the bird's brachial vein with a sterile syringe needle to collect a few drops of blood. (h) Eventually, he stores the blood taken from the bird. (i) Holding the bird in his right hand, he takes further notes with his left, (j) documenting the measurements of the bird. (k) Next, he measures the bill length, (l) tail length, and (m) tarsus length with a calliper. (n) The last measurement is the bird's weight with a scale. (o) Michael notes down all the results. (p) At the end of collecting data, the bird gets rewarded with its 'favourite dish', sausage, before it is released. (q) Before continuing with fieldwork, Michael attaches the tail feathers he removed into his field notebook with tape, next to his notes on the bodily data about the Siberian jays.

Figure 37:
Box with blank and pre-written 'ringing sheets' as they are stored in the field house. The biologists usually take a few sheets from this box into the field to register new Siberian jays. Arvidsjaur, 2015.

Figure 38 a–e:
Sequence of processing the raw data collected in the field. Arvidsjaur, 2015. (a) Michael handling the day's blood samples taken by Julian, Kate, and himself, on the kitchen table. A field notebook in the background and cutting board to prepare dinner. (b) Michael labelling the blood samples by noting the birds' IDs on tape with pencil and sticking them onto the samples to make sure to keep the overview, not mix them up, and maintain consistent data. (c) Other field material is lying on the table. (d) They stick the tail feathers they have brought back from fieldwork into a notebook and mark them with the relevant bird IDs. (e) Finally, the blood samples from the Siberian jays are stored in a tin can on the freezer shelf of the fridge.

4.3. Situated Enskillment

Biologists conducting fieldwork usually have different levels of knowledge of the study. There are very experienced biologists, such as Michael, less experienced biologists, such as Julian, who has been to the field a few times, some who are experienced field assistants but in other studies, such as Marine and Camille, and then complete novices, such as Kate. All researchers in the study, except for me, are formally trained biologists from different universities, who have learnt the practices and theory of biological research. While, on the one hand, they are familiar with propositional and formal biological knowledge, their education also gives them access to the relevant professional network that brought them to Arvidsjaur. Kate and Julian met Michael through the university as his students, whereas Marine and Camille were recommended to him by a former field assistant. Their ability to work as biologists and do biological fieldwork is certified by an academic degree, such as a BSc, MSc or PhD in the field of biology. This degree officially makes them peers and insiders of the discipline with a specific, practical skillset that engages bodily, sensory, and propositional theoretical knowledge. They are aware of current research methods, questions, and practices of analysis, as well as the rhetoric of the natural sciences and the overall practices of knowledge production. They, at least theoretically, know how to conduct biological research.

However, formal training and theoretical knowledge are not sufficient; their education as biologists does not help them to conduct their observations and experiments in Arvidsjaur with the Siberian jays. They need to undergo a further situated enskillment, as I shall call it, that meets the criteria and requirements of the Siberian Jay Project. This *situated enskillment,* inspired by Haraway's situated knowledge[13] occurs in situ during fieldwork. While the

13
Haraway, 'Situated Knowledges'.

body is always part of a situatedness, the term 'situated' emphasises the relation to a specific environment. In this case, the main aspects are the scientific context and the forests. In this sense, *nature* and *culture* come together in a way that co-constitutes them, making it difficult for them to be divided. The specific field situation needs formal knowledge from university training to be transferred into practice, thus requiring enskillment to this specific study. It is highly dependent on the location in which the biologists conduct fieldwork with the Siberian jays and the specific practices of data collection, as performed by Michael and his team.

Before participants who are new to the study – novices – can collect data on their own, they must undergo a process of enskillment. This enskillment serves two purposes: First, it enables the novices to acquire the skills of cultural and bodily practices[14] that facilitate data collection, such as handling the tools and devices, knowing how to back up data, and how to operate during fieldwork. They may also involve practices of drawing, writing, and note-taking, as well as digital and geographical literacy. Second, it enables them to acquire the professional skills necessary for data collection, which involve knowing what to look for, how to interpret behaviour, and the methods to formally document this. Data collection is possible only based on a combination of the two, and to apply the correct practices, the novices need to undergo training. This training, again, can be divided between the formal training received at university to become an educated biologist and learn how to conduct biological research, and the situated enskillment required for this specific study in the field. According to Michael, fieldwork is 'highly creative, requires improvisation and the ability to work independently' and these are aspects that cannot be formally learnt in university halls but only through immersion in the field.

14
Mauss, 'Techniques of the Body'; Maye, 'Was Ist Eine Kulturtechnik?'

In 2015, Kate had never done fieldwork or collected any data on birds. Thus, she was a complete novice to the study. While the two field assistants, Camille and Marine, whom I accompanied in 2020 were new to this study, they were placed on Kerguelen for the French Antarctic Program where they had gained ample field experience. And then there was me, a complete novice and a disciplinary outsider. In 2015, I had not undergone any enskillment in biological fieldwork; however, as a former designer I was familiar with the cultural practices they employ during data collection, such as visualisation.

Kate and I were the novices in 2015 who had to undergo situated enskillment. In contrast, in 2020 Marine and Camille were the novices. While for Kate, Camille, and Marine, situated enskillment was part of their job as field biologists, for me, it was part of my methodological approach. My aim was to become familiar with the sensory and bodily practices that enable knowledge production through participant observation and a sensory ethnography. Accordingly, I also underwent an abridged version of situated enskillment, mostly by assisting the biologists and witnessing the processes of enskillment. By accompanying this process of enskillment, I gained an understanding of the processes of data collection and the visual practices the biologists employ.

As part of the situated enskillment, the biologists learn what educational historians Thyssen and Grosvenor call the 'modes of meaning-making',[15] as they occur in this case study: 'Meaning-making and the modes employed to bring it about refer to cultural practices such as seeing, reading and writing, which in turn involve the handling of things and artefacts'.[16] Meaning-making does not occur only with

15
Geert Thyssen and Ian Grosvenor, 'Learning to Make Sense: Interdisciplinary Perspectives on Sensory Education and Embodied Enculturation', *The Senses and Society* 14, no. 2 (2019): 120, https://doi.org/10.1080/17458927.2019.1621487.

16
Geert Thyssen and Karin Priem, 'Mobilising Meaning: Multimodality, Translocation, Technology and Heritage', *Paedagogica Historica* 49, no. 6 (2013): 736, https://doi.org/10.1080/00309230.2013.848912.

fact-based knowledge, but needs to be complemented with a sensory enskillment involving 'the visual, aural, audio-visual, material, bodily and spatial'.[17] They refer to these capacities as 'embodied enculturation' arising from sensory enskillment, whereas I refer to it as situated enskillment to emphasise the role and meaning of the context in which the enskillment takes place.

This enskillment involves wayfinding practices, the practices of registering new birds, data collection, and observation; and learning about the technologies used in documentation practices. Most of these practices are accompanied by media and tools that extend and/or complement the bodily capacities and organise the data collection. Accordingly, their employment and handling must also be learnt, which I conceptualise as *situated mediations* in practices of *Collecting* (Chapter 5).

One of the first aspects to learn as novice is the reference system, which involves epistemic qualities that help biologists structure and organise the fieldwork. Only through these references – immutable mobiles – can data be collected consistently[18] and retraced back to the birds once they have been translated into a code, based on further technical manipulations (as discussed in Chapter 6 of the book). The reference system brings me back to the field notebook and the datasheets and maps contained therein. Knowing the reference system means being able to relate the Excel data plots and maps to their living referents in the forests, the colour-ringed Siberian jays. The ability to make these connections and find the actual locations of the bird territories in the field based on a GPS and map are the most basic skills required to conduct fieldwork and formally collect data.

Novices must also become familiar with the methods of data collection in this study (Chapter 5). While the general

17
Thyssen and Grosvenor, 'Learning to Make Sense', 120.
18
Latour and Woolgar, *Laboratory Life*.

method is a standard one of observation and documentation in behavioural ecology,[19] certain boundaries and grey areas that are not predefined must be agreed on by the biologists. This could be the frequency with which a behavioural activity is documented, for example, whether the biologists count the same behaviour that occurs twice in a row without anything else changing as one long activity or as two separate activities, and what behavioural patterns will be documented.

The biologists must learn how to set up the observations, what to focus on, and how to document it. The documentation method is particularly important because it defines what can become data and how. It determines the medial form and the quality and kind of data, and only if this method is employed consistently by all biologists does it allow for the merging of individual data points into one data plot where they can then be quantitatively analysed and compared (Figure 23).

The biologists also need to know *how* to record observations, which methods and notation techniques are used, and *what* to record, which is knowing what to look for. Lastly, the novices must learn precisely how experiments are conducted in the circumstances in Arvidsjaur. Therefore, all new study participants, regardless of their prior experience as field biologists, must receive specific situated training. Their university training will have focused on the technical aspects and theory around fieldwork, but not the actual practices of observation and data collection.

Situated enskillment is based on cooperation with other biologists and peer learning, with a protocol established for data collection. Within a student–master relationship, the novices learn by accompanying biologists who are already *in the know*, shadowing them for the initial days of research as assistants rather than independent biologists. Through this approach, they become familiar with the methods of data collection and documentation. They

19
Martin and Bateson, *Measuring Behaviour. An Introductory Guide.*

learn how the field is structured, how to navigate it, how to identify birds, and how to do the observations – which involves knowing what to look for and pay attention to, what to ignore, how to document these observations in their notebooks, and how to ring new birds. As I have witnessed and will show, these initiation processes are passed on between peers and are not based on textbook learning. They are developed through a communal learning process and social apprenticeship.

Thus, I received different perspectives on the apprenticeship and the required formal and informal/implicit skills. They became apparent to me when accompanying the individual biologists during their enskillment. Not only did I observe how they learn and collect data but I also documented the individual moods, mindsets, sensory attunement, emotions, and aesthetic thinking they engage in during fieldwork. Aspects such as these become particularly visible when examining scientific practices from a design perspective, which returns to the scientific process what is usually eliminated. In addition, the situated enskillment is also significantly bound to the working environment, namely the field and the conditions, which only allow certain practices. Thus, I highlight agencies that become invisible through the translation process – from field, to office, to paper – and aim to make them visible again for a more in-depth understanding of the practices of preparation, collection, and production in evolutionary biology.

4.3.1. Wayfinding and Working with Maps

The first steps of fieldwork, before the biologists can collect data, are shaped by finding the individual bird territories in the study areas. This is a complex task executed differently by each biologist. The study area is defined according to the bird territories and the way the birds inhabit the forest. Studying the maps that the biologists use to maintain an overview of their study reveals that the territories are rarely located along streets or trails but rather deep into the forest. Usually, the birds have selected sheltered locations where they feel safe from predators, can build nests, and can find and hide food. The individual territories are not always easily accessible and require the researchers to venture away from the human-made paths. Thus, new biologists must first establish their wayfinding skills.

Finding paths can be difficult, especially in winter, when the entire landscape is covered in a thick layer of snow. For example, small bridges may have become invisible apart from a tiny mark on a map, and distinguishing them from snow that has been bulked up across the half-frozen river by the wind can be challenging (Figure 24). While, most of the time, it is fairly safe to trust one's visual judgement when navigating the field, moments such as this require additional navigation tools. Equally, wayfinding by means of maps is also unsuccessful in these conditions, as the references, such as tracks, rivers, or marshland, that support navigation have often disappeared under the snow or disappear once one has veered off the main paths and entered the forest.

Depending on their experience with the environment, the biologists use a combination of maps, tools, and techniques to navigate the study areas, where there are usually no signs or markings to guide the way; nor can they rely on smartphones for navigation because they run the risk of a flat battery or lack of signal. Without reliable devices and knowledge of the landscape, new biologists would easily become lost in the study area. Thus, wayfinding, based on spatial and visual orientation during fieldwork, is a vital skill that must be developed first.

In the anthropology discourse, wayfinding in the sense of spatial cognition is often and quite generally separated into route knowledge and mental maps. Route knowledge, also referred to as 'practical mastery', 'permit[s] generalizations and inferences beyond the specific spatial information gained through direct experience'.[20] Mental maps, which are also associated with 'survey knowledge', work as 'cognitive representations of the spacial relations between objects',[21] and are thus 'survey, configurational, or layout knowledge as opposed to route or sequential knowledge'.[22] The mental map is also characterised by the ability to make shortcuts, detours, and point out relevant landmarks and 'the gradually changing vistas along the route'.[23] This knowledge is based on experience with the environment, which 'enables a subject to find a novel and more economical route [...] the ability would be absent if he/she had only route knowledge'.[24] Thus, it is understood as 'the process of moving from one recognized visual perspective (vista) to another (transition between vistas)'.[25] In this sense, when new biologists join the study, they must first develop their wayfinding practices and the spatial cognition required to navigate the field.

The novice becomes skilled, not through the acquisition of rules and representations, but at the point where he or she is able to dispense with them. They are like the map of an unfamiliar territory, which can be discarded once you have learned to attend to features of the landscape, and

20
Gary Allen, 'Spatial Abilities, Cognitive Maps, and Wayfinding: Bases for Individual Differences in Spatial Cognition and Behavior'. In *Wayfinding Behavior: Cognitive Mapping and Other Spatial Processes,* edited by Reginald George Golledge, 46–80 (Baltimore: Johns Hopkins University Press, 1999); as cited in Istomin and Dwyer, 2009a, 32.

21
Kirill V. Istomin and Mark J. Dwyer, 'Finding the Way: A Critical Discussion of Anthropological Theories of Human Spatial Orientation with Reference to Reindeer Herders of Northeastern Europe and Western Siberia', *Current Anthropology* 50, no. 1 (2009): 29.

22
Ibid.

23
Tim Ingold, *The Perception of the Environment: Essays on Livelihood, Dwelling and Skill* (Taylor & Francis, 2000), 191.

24
Istomin and Dwyer, 'Finding the Way', 32.

25
Ibid., 29.

can place yourself in relation to them. The map can be a help in beginning to know the country, but the aim is to learn the country, not the map.[26]

Physical maps and navigation tools may be used to obtain an overview. In my case study during the first few days, they are essential to navigate the landscape. However, 'to learn the country', and develop a mental map and route knowledge, the biologists slowly have to evolve from studying the GPS for guidance to observing the landscape and memorising landmarks as reference points to navigate between bird territories.

During wayfinding, the biologists combine route knowledge with survey knowledge, which they gain from maps. Finding the way requires a certain geographical literacy[27] and spatial cognition. The most frequent combination of navigation tools I observed is a combination of a GPS and a map of the three study areas, *Managed, Reivo,* and *Fat Road*, as they are glued into the field notebooks (Figure 16).

These maps are highly formalised and offer only an approximate representation of the actual territories. They do not help the biologists navigate, unless they have prior experience with the study area and have already developed a mental map that they can follow to some extent. The schematic maps of the study areas by themselves are not meant to provide orientation but rather to help plan the study and maintain an overview of the progress made in the individual territories. To compensate for their shortcomings, they are usually employed in conjunction with a GPS. Locations where other biologists found birds' nests

26
Tim Ingold, *The Perception of the Environment: Essays on Livelihood, Dwelling and Skill* (New York: Routledge, 2000), 415.

27
'The ability to understand, process, and utilize spatial data is termed geographic literacy, a skill that plays a significant role in our everyday lives whether or not we are geographers'. Sally Turner and Joseph Leydon, 'Improving Geographic Literacy among First-Year Undergraduate Students: Testing the Effectiveness of Online Quizzes', *Journal of Geography* 111, no. 2 (2012): 54, https://doi.org/10.1080/00221341.2011.583263. It involves 'geographic knowledge and geospatial recognition'. Thus, it encompasses the ability to employ geographical language and visual and spatial skills.

the previous year are marked with emojis in the GPS, for example 🍺 (a beer jug emoji). This is usually also the spot where the biologists want to work with the birds.

All the biologists have access to the same digital maps of the correct locations where they collect data every year, even if their individual mental maps and orientation change. Thus, they can all share, use, and edit the information. The topography of the site is also represented on the GPS maps, which helps in difficult situations, for example, when fieldworkers must cross snow-covered rivers or find a specific path. However, a GPS and geographical literacy alone are insufficient. A greater sensory attunement to the landscape is necessary for fieldwork. The biologists need to 'learn the country', and develop their own spatial orientation, which works somewhat independently from additional tools. Only once the biologists can stop studying the tools for guidance can they attune to the environment and navigate efficiently, and thus focus on the birds.

Michael has acquired extensive survey knowledge based on mental maps and routes, and is aware of the importance of this geographical overview. He knows that orientation of the study site is vital for navigating the field successfully and not being distracted by tools and becoming lost. A mental map also helps to identify and imagine locations on the maps (landmarks) more easily, while the combination of mental survey knowledge and maps facilitates phone conversations during fieldwork or when planning in the mornings and evenings. Thus, Michael uses the opportunity to help the team develop their survey knowledge and mental maps whenever we are driving in the field car, passing parts of the study area. He repeatedly points out where the study area starts by saying 'here, where you can see this treeline, is where our study area begins', drawing an invisible line along the trees ahead of us with his hand, which is divided by the street from which we are about to enter the study area: 'our study area'. With this, he relates the territory to the surrounding environment. This oral and visual marking of the area is followed by pointing out specific territories along the street, naming them, and – often on my request – telling an anecdote about them, usually those that explain the territory names.

We drive past the village Akkavare, where Lindgren, the primary school teacher who observed the jays until he died in 2017 at the age of almost 100 years, grew up. Michael points out the wooden cottage where he grew up, honouring his work. He also draws our attention to the bird territories around the village where Lindgren undertook his research: 'Folke's birds'. Michael does this mapping task almost every morning while we drive. As soon as we enter the study site, he informs us that we have now crossed the (imaginary) border.

These small sequences of enskillment are often accompanied by comments on how the forest management has advanced and how it has influenced the study area; thus also highlighting the changing vistas and the problem that the spatial orientation may be affected if vistas disappear with this transformation of the environment. This eliminates the connection between the vistas and the biologists' geographical survey knowledge, as their reference points have disappeared. Again, Michael points in their direction while driving, marking geographical locations that his peers may add to their spatial knowledge and that together may constitute their survey knowledge and mental maps. Despite his knowledge of the area and the local politics of Arvidsjaur, he occasionally seems surprised that the forest is gone. He provides background on the political and economic situation in Sweden and how landowners often decide to fell their trees for money without thinking of the consequences, despite a low market price for wood (although it has more than doubled since 2020).[28]

28
While in 2020 the prices for wood were still rather low, since that time the prices have risen by at least 50–70%. According to Michael, this has partly positive short-term impacts on deforestation. In particular, the private owners of the forests see them as a long-term investment, and have reduced deforestation because they expect the prices to rise even further. However, the state forest company, Sveaskog, continues cutting down the forests mostly to meet the high demand for paper, according to Michael (cf. zoom interview, 23 May 2022).

What the entrepreneurs consider a commodity is, for Michael, the habitat of his birds – something that requires protection for his study to continue, in addition to his broader commitment to preserving the environment and local ecology, shared by his scientific peers. The conflict between Michael's work, the way he perceives and uses the forest, and the entrepreneurs becomes visible during these short car journeys and will accompany our fieldwork, particularly in the 2020 study.

Wayfinding with Michael, Marine, and Camille

On the first day of fieldwork in 2020, Michael, Marine, Camille, and I drive to the study area, *Managed,* to put into action what had been planned and discussed at the kitchen table that morning. Michael parks the car near the best access point to the bird territories and we unpack the wooden touring skis we use to move through the forest, secure our leather boots in the metal binding, and make our way towards the bird territories.

The four of us ski along a forest road towards the first territory we shall observe this season, *Kara.* On arrival, Michael starts calling for the birds by making whistling sounds and attaching a feeder on a tree. Marine and Camille, who have now entered the mode of situated enskillment, focus on what Michael is doing. He first gives advice on how to study the birds. It appears to me that his training method is not a didactic step-by-step guide, but rather one of 'learning on the go'. He comments on everything he does and explains why he does it, occasionally sharing advice on how to approach the field in general. Marine and Camille stand next to each other, a short distance from Michael, and start imitating what he is doing; looking at the birds that now are starting to arrive through their binoculars and making their first notes.

Standing here in the low sunlight of Sweden's winter sun with Michael, Marine, and Camille in Spring 2020, I am reminded of my first field day in 2015. Michael had invited me to accompany him for fieldwork, while Julian was training Kate. Once we had dropped them off and arrived at the place where we would enter the study area by car, we put on our skis and our backpacks filled with our equipment. Back then, I was a complete novice. Not only did I barely know what would be expected of me during fieldwork, I had also never used the kinds of skis that we were using to move through the field. Once I locked my slightly-too-big leather boots into the clunky metal bindings, we started skiing into the forest. I vividly remember how, without any advice on how to use the skis, Michael set off at speed into the forest, following his spatial knowledge and disappearing before my eyes. Despite being a good skier, I was struggling to keep up with him. This was the moment I realised that this was not going to be an easy trip. It was going to be work – fieldwork – and Michael had just entered the field. From now on, he was in fieldwork and data-collection mode. This had top priority and, as with everything else, I had to align with it.

Michael suddenly changed direction, leaving the small path we were following to head deeper into the forest. He made this decision based on route knowledge, probably vistas, alone. He then slowed his pace and suddenly stopped. As he hung up the fat on a suitable tree (Figure 25), he started calling the birds and looking through his binoculars. After a few moments of whistling, the first birds arrived. He identified the first bird as female, looking through his binoculars and studying the coloured rings on its legs. He commented that the bird was still timid, and he identified the bird sounds in the background as that of willow tits that also live in the forest. Sometimes we would hear the calls of owls or ravens, both predators of the Siberian jays and a possible reason for their absence. We continued waiting until more birds arrived. Then, once the entire group of birds was present, Michael began formally observing their behaviour. He commented that they were mobbing one another and that he was now focusing on their social behaviour within the group. Back then, he also checked his GPS to see whether the feeding spot was marked on the map.

This scenario surprised me the first time because I could see no indicators of why we had stopped at that exact location. Thus far, I had imagined that we would walk through the forest until we saw some birds, whereafter we would study them through our binoculars. However, Michael and his team followed an entirely different process to what I had imagined. There was a specific procedure with clear aims on where to go, and we never wandered aimlessly through the forest looking for birds. It was the other way around: we would go to where the birds should be, attract them, and then study them. There was no 'searching' for the birds, but rather 'calling' or 'checking for' them, as Michael refers to it. He attracted the birds to where we were, and, as I learnt later, Michael knew from experience that this small spot between the trees, away from the paths, where he had decided to stop, was the centre of the first bird territory that we visited that day. It would be where the nest would be located in summer and where we would now set up the fat to attract them (Figure 26).

Now, in 2020, I am again standing in this cold but beautiful scenery of Arvidsjaur, and it is the first field day. Being surrounded by the landscape, sounds, sights, birds, snow, and trees, and with my feet attached to the skis, evokes memories of my previous field trips. Now, Camille and Marine must become enskilled, and Michael suggests that one goes with him and one with me, as I 'half know the deal', according to Michael. Before we split up, Michael explains to Camille how to get to the territories, named *Troll* and *Måskomyran*, where we shall meet later. They are standing close together, studying one of the small maps of the study area that they have glued into their notebooks, as I had done with Julian at the kitchen table before my previous field trip with them. Michael does not use the maps for guidance, but rather to show Camille where to go. He names the territories, points to their location on the map, and describes the route. The way he explains the route to us reveals his route knowledge, as he points out specific vistas, details on the path, and landmarks that help with orientation.

In this way, he not only describes the route, but is also forced to recall the route by heart and thus update and actualise his own mental map and route knowledge; and, as today is the first day of fieldwork, probably also activate it. Over the next few days, Michael will also pass on his perception of the environment and the aspects to which he pays attention by pointing out specific vistas and landmarks when describing the routes to Camille and Marine. Furthermore, as they attempt to follow his directions, Marine and Camille develop mental maps based on the same indicators. In this way, route knowledge is passed on between peers, thus contributing to the situated enskillment of the fieldwork in Arvidsjaur.

I am not engaged much in this process; my duties are limited to helping Camille with the more specific practices of attracting Siberian jays, such as setting up the observations and documenting them. Accordingly, Camille is in charge of the route, as I have only a basic idea based on the directions Michael gave us before we walked off into the forest by ourselves. Camille and I fail to attract birds in *Troll,* where we navigate to with the additional help of a GPS (Figure 27). We return to *Kara* and swap teams, where I go to the next territory with Marine. On our way, we stop and whistle after what felt like 900 m, which Michael had asked us to do on a phone call. A bird crosses our paths, flying above our heads. We identify it as a Siberian jay and note that it seems to be wearing coloured ring markings (IDs). We keep calling for it, and Marine tries to trace it through the binoculars. However, it flies away and she does not manage to identify it. When we arrive in *Måskomyran,* where we all plan to meet again, we can only tell Michael that we saw a Siberian jay, not which one.

Marine

The next day, after we have headed to a few territories together, I set off with Marine. We follow Michael's instructions on getting to *Rågången*. We ski down a road until we arrive approximately where Michael told us to go. Now, Marine's mental map that Michael had started to develop with her the day before is insufficient and she takes out the GPS for more specific indications of the route. Before we left, she had typed in *Rågången* to get the directions. She checks the screen while I wait beside her, and then decides on a direction.

According to the GPS, we must follow a straight line into the forest for a few hundred metres. However, how does one ski in a straight line between bushes and trees without a horizon to focus on? Once we have skied a few hundred metres into the forest, we are lost and must completely rely on the GPS. Navigating with a GPS, while also being new to moving on skis that require both arms to use the poles and maintain balance, while also pushing oneself forward, proves to be difficult. I follow Marine as she is engaged in guiding with her entire body. Ultimately, we ski around the bird territory twice, in circles, as the tracks in the snow brutally reveal to us.

This wayfinding cost us time, which discourages Marine. She becomes frustrated and comments, 'Oh, tomorrow I will bring a compass! This is so inefficient with the GPS. I am used to maps and compass'. Moreover, the GPS works only when moving, and it has a small delay, which is problematic when skiing. However, it is sufficient if we follow a straight line. The GPS also turns off when it is not being used, and every time Marine wants to check the screen, she has to start the device again and wait for it to turn on and update the location, which costs us a significant amount of time. Finally, the arrow on the GPS gets closer to the waypoint that marks our destination.

Once the arrow representing Marine's location and the waypoint of the territory overlap on the GPS, Marine and I set ourselves up for the process of behavioural observation (Figure 28). We drop our bags, take a brief break, and take out the equipment we need. Now Marine finally has some time to share her thoughts with me. She explains that 'the first days, it's always difficult, I have to get used to everything here first: the gear, the way I move, the weather, the codes for the behavioural observation. It takes a few days but I'll get the hang of it'. While she clearly has the professional knowledge, as she has done much fieldwork in the past, what she describes to me is her lack of the situated enskillment, which she only started acquiring today.

Situated enskillment involves the formal knowledge of data collection, and the knowledge of navigating the field on skis while consulting a GPS. It also includes a sensory attunement to the landscape to understand where one can ski, and attunement to the birds. For instance, when it comes to determining where to set up for the observation, route knowledge is required, which results only from training, enculturation, and situated sensory enskillment. It also involves being able to read and identify clues in the landscape that help develop a mental map and improve navigation in the field.

En route to the next territory, which no longer follows any streets or tracks, Marine changes her navigation strategy once we are outside the dense forest and on more open marshland. She checks the GPS and then chooses a general direction and a spot on the horizon; this is a skill she must have transferred from other experiences of moving through natural landscapes without navigational help. Even though it seems difficult to move in straight lines on the skis, we manage to reach the next forest line fairly directly in this way. Once we enter the forest, wayfinding becomes difficult again. Marine takes out the GPS and I follow her. Walking behind her, I can tell that she is struggling with learning to ski and simultaneously guiding us through the unknown landscape. She skis with one pole in her right hand to push herself forward, while the other pole dangles from her left hand, in which she is holding the GPS, drawing thin lines in the snow.

Guiding us closer to *Nadine,* the territory where we are going to meet Michael and Camille again, Marine's gaze switches between the screen of the GPS and the landscape, as she must be careful about where she goes next, avoiding walking into bushes, falling into rivers, or colliding with trees. She continuously shifts her attention between the GPS and the environment, sometimes navigating difficult terrain because the GPS is not always an exact representation of the territory. However, to collect information from the immediate environment, she must develop her own spatial cognition for this study area. Her solution of combining the GPS for general direction with landmarks and vistas as further guidance appears to be a smart one.

The way the GPS is designed makes it appear to represent every tree, bush, fence, and river on its maps, so that one can move safely and efficiently through the territory by simply consulting the map and following the arrow. However, the GPS is not that precise and is not always updated. We discover that the GPS is of no help when we want to cross a river because it lacks this precision. We then return to our visual judgement without any technological assistance and decide that the snow pile a few metres further down the river is most likely the bridge. We manage to cross it successfully, uncertain of whether the pile of snow we crossed on skis was actually the bridge we had been looking for, or if perhaps it was just a tree that had fallen over the river, or some ice and snow that had turned into something resembling a bridge, a similar experience to one I had with Kate in 2015.

Interestingly, this bridge has inscribed itself as a landmark in my route knowledge of the study area as we spend significant time studying the environment, thereby cementing it as a vista. This was necessary to gauge whether we were taking the correct and safest route. These moments of struggle are potentially decisive in developing spatial cognition. After having crossed the bridge, we finally see Michael and Camille waiting for us in *Nadine*.

Camille

In *Nadine*, we swap teams again and I accompany Camille. Having been to a few territories with Michael and on her own, she already seems more dexterous with the equipment than she did the previous day. Now she moves through the forest in a determined and self-confident way. After we left Marine and Michael, where we repeated all the steps of fieldwork together, we make our way to *Södra G*. We arrive via a small detour because we walked on the wrong side of what appeared to be a snowmobile path. We alert the birds to our presence with some whistling on arrival. Just before we stop, Camille double-checks the GPS to make sure that we found the correct spot. She then stops, removes the fat from her bag, and puts it up, while I set up a video camera with which she will record the behaviour of the birds.

After we have finished setting up, we stand next to each other silently, as we do on many more occasions over the following days. Camille is looking at the trees through her binoculars, while occasionally breaking the silence by whistling.

I am taking notes in my field notebook to document what she is doing, the set-up of this territory, the weather conditions, and my thoughts. After a few minutes of whistling, which I occasionally join, and just after our first thoughts of leaving the territory, thinking that the birds will not show up, the forest is suddenly filled with life. Between the tops of the trees, the birds invade our surroundings one after the other from all directions, diving towards the fat Camille has put up. Once they start feeding, the birds settle, peck at the fat, fly away, and return a second later; they dismiss other birds, revealing their hierarchies. Flapping their wings, they twitter, and jump between the snow-covered ground, the fat, and branches of the trees.

Sometimes, this spectacle takes a rapid turn when birds from nearby territories arrive. Then, the scene turns into what Michael calls a 'fight'. Rather than all eating from the feeder together, the birds banish one another from it. This lively scene means that the biologists cannot use the data because they are not studying the competition between bird families. Despite this unplanned situation, Camille starts checking the colour-coded rings of the birds, notes them, and checks if all the birds of the group(s) are present. This at least helps to obtain data on the bird population and their cooperative behaviour. Most of the time, however, only one group appears; 'fights' are rare and occur only once every few days. If the correct group emerges for the territory, according to the data plots in Camille's field notebook, she starts observing them in a more formalised way based on the protocols for behavioural observation that Michael had introduced her to during the last couple of days. Once the relevant data have been collected, which I describe thoroughly in *Collecting* (Chapter 5), we continue our fieldwork approximately 30 minutes later and move on to the next territories.

This next territory, *G-stjärn,* is close to a snowmobile path where the snow has been flattened into a small road that serves as a landmark we can follow without much additional navigation. It is thus easy to find and follow. Now, the forest becomes less dense. On arrival in *G-stjärn*, we manage to find the tree on which the biologists from the previous year must have put up their fat. The wire is still attached to the tree, which becomes an unintentional landmark and affirms that we have arrived at the correct location.

To both of us, this seems a rather unusual territory for birds because it is somewhat unprotected and right at the edge of the forest. Nonetheless, we start whistling to the birds. We both feel confident that we are at the correct spot, as the wire on the tree reassures us, more so than in other locations. We continue whistling and look up into the trees, Camille with the help of binoculars and I with my bare eyes. Then we take a break from whistling, let the silence sink in for a moment, exchange a glance and then start calling to the birds again by imitating their noises. It takes a long time; the sky is grey, and the missing trees create a bleak scene.

The conflict between the biologists and the landowners who have felled the old, tall pine trees for profit becomes visible here. It appears we are standing right at the boundary where it is not quite clear whether the forest with the bird territories will remain, whether it has already disappeared, or whether the trees that form the edge will be felled next. The birds do not appear today, even though we are positive that we are right in the centre of the territory marked by the wire – Camille double-checks on her GPS and concludes that the birds may have changed territory owing to the invasion of their habitat.

This conflict is not merely one between biologists and landowners. The sight is also reminiscent of a historical conflict between landowners and the Sámi people. It had long been disputed whether landowners have logging rights or whether the Sámi were granted reindeer grazing rights in several areas of the Sápmi region. Only in 2011 did the Supreme Court decide in favour of the Sámi for the region of Västerbotten,[29] which lies around 50 km south of Arvidsjaur. Wood is not the only raw material of value in the land of the Sámi; other critical materials are mined. Following the Critical Raw Materials Act of 2023, so-called 'essential mineral supplies crucial for the EU's green and digital transitions'[30] are extracted here. Moreover, it is also a popular region for governments and industry to develop hydro and wind power.[31] As such, the Sámi are suffering not only from the impacts of climate change on their lives, which are shaped by Indigenous practices, but also from green colonialism. Let us return to the forests of Arvidsjaur and the Siberian jays.

On our way to the next territory, *G-liden,* the impact of deforestation becomes even more extreme. What is marked as a dense forest on the GPS reveals itself to us as vast, open land. Where we expected to encounter trees and groups of birds, we find ourselves standing in a wide, open area covered in snow, with a few individual trees left in the centre, almost as a reminder of what once used to be there (Figure 29). The GPS indicates a dense forest, but this is not the reality. Here, the vista has changed; and while Camille does not know any different, as she is here for the first time, Michael will have to update his route knowledge when he arrives. While Camille does not require any *updating,* as she must first develop a mental representation of the landscape, Michael's existing memory of the landscape will be overwritten by its new version soon. This will not be in such a way that his memory will be erased entirely, but rather so that he becomes a witness to the

29
Eivind Torp, 'The Legal Basis of Sami Reindeer Herding Rights in Sweden', *Arctic Review on Law and Politics,* 1, no. 4 (2013): 43–61.

30
The European Parliament website, 12 December 2023 https://single-market-economy.ec.europa.eu/sectors/raw-materials/areas-specific-interest/critical-raw-materials/critical-raw-materials-act_en.

31
David Nathaniel Berger et al., '2 IWGIA – The Indigenous World – 2024', 2024, 463–70.

'temporality of the landscape'[32] and its permanent transformation, knowing the past and the present appearance of this area. Through forest management, this landscape becomes increasingly inscribed with anthropogenic traces. This observation, again, will enhance the biologists' implicit knowledge of the study: making them experts in data collection from the birds and the birds' environment. This is not the only instance where some territories have disappeared owing to forest management.

When we share our observation with Michael later that day, he is surprised to hear that this section of the forest has been all but eradicated. I wonder how this will influence his geographical knowledge of the study area if such fundamental changes occur from one season to the next. In situations such as these, with all reference points having disappeared, Michael's oral descriptions no longer suffice, and Camille must revert to following the GPS.

It is easy to change direction and ski straight towards the next territory given that there are no trees in our way. In fact, I enjoy the way the skis carry us across the untouched snow without any of the usual obstacles. After a few hundred metres, Camille stops, checks the GPS, adjusts her direction slightly and continues towards the next patch of trees and the next bird territory to which Michael had directed us. Then, she receives a phone call from Michael asking us to come to the nearest street where he will pick us up. Again, Camille looks at the GPS, chooses a direction, and starts skiing with me following close behind. Cutting a straight line into the snow, we soon arrive at the main road, take the skis off, and wait for Michael to appear in the field car.

If one were to take an aerial photograph of our tour today, the lines we had drawn in the snow would reveal the network of bird territories, the trace of our day's fieldwork. The lines would also reveal that, aside from the first day with Marine, we did not have to navigate through difficult

32
Tim Ingold, 'The Temporality of the Landscape', *World Archaeology* 25, no. 2 (1993): 152–74.

territory and mostly followed straight lines: first on the snowmobile path and later through the bare forest. We drew straight lines because there were no obstacles or trees to navigate around, given the forest was gone and, with it, some bird territories. These traces will eventually also change the biologists' wayfinding behaviour, as they offer new, temporary reference points for orientation and navigation.

Michael

Today we drive to *Reivo,* a nature reserve where trees cannot be felled; for this reason, wayfinding is different here. The paths are more challenging but the landscape is more unique, and landmarks and vistas are easier to identify. According to Michael, *Reivo* has 'the oldest birds, as the best forests are here'. The car journey to *Reivo* is longer. After around 20 minutes, we enter *Managed,* 'our study site', and pass Lindgren's house. After approximately 40 minutes, Michael parks the car next to a shed in a small village consisting of one house – nowadays inhabited only on weekends because the owners work during the week along the coast further south – and a farm. We have arrived in *Reivo* (Figure 30).

In *Reivo,* we unpack our material, pull the lever to secure our leather boots into the ski bindings, and, with Michael leading the way, enter the study site as a group one last time. After a small ski tour, following a path through the dense and beautiful winter forest of *Reivo,* Michael stops. He does not double-check but seems to know that we have arrived at the bird territory, *Maderängen,* which is near the path, thus offering a prominent landmark. Together, the biologists begin attracting the birds through a rich pattern of calls. The jays arrive quickly, which is often the case when Michael is around. Over the next 15 minutes, he again shows Marine and Camille how to record their observations. The three of them stand beside one another, and while Michael installs the camera, Marine and Camille begin observing the birds, switching their gaze between their notebooks and the birds – with the help of binoculars, this practice has become much more routine now. Michael explains how the camera should be positioned to film the birds and what should be in the frame (Figure 31). He then comments, 'I will break the branch off', finishes his set-up, and walks towards the tree that is in the way.

Once this is done, we discuss who will collect data in which territory (Figure 32 a). The three biologists have their GPS devices on and notebooks open, ready to take notes. Michael defines clusters of territories that could be visited together, also describing the general characteristics of the route and the levels of difficulty and skiing effort it will take to get there. Again, he mostly speaks from memory rather than using a map with a geographical profile. He occasionally checks the map of the study site and uses it to show Marine and Camille where the territories are located. The clusters contain four to five territories each with some additional options if someone is fast. Those who enjoy skiing go to those that are farther away, while the others remain near the car. Marine heads to the easier territories, while Camille heads to the more difficult ones. Michael explains to Camille that *Norna* is very difficult to reach and shows her the directions on the GPS. He also mentions that if she does not want to go (Figure 32 b), Julian might be able to do this when he joins them in a few days, once Michael and I have left, as 'he likes skiing'.

In that moment, I remember the *Norna* study site, which I had gone to with Kate in 2015. It was one of our worst experiences. I was following Kate, who, without any route knowledge, had chosen an arbitrary path up the steep hill, simply following the GPS in a straight line. She did not pay attention to the properties of the landscape indicated on the map because of a lack of geographical literacy, such as reading the map and the indications for altitude. While the route up was very steep and exhausting, coming back down was dangerous. I remember commenting on my bewilderment that Michael had sent us up there. I had drawn this route with a zigzag line into my field notebook with the comments 'up the hill, no birds, come up all the way for nothing'.

As for today, I decide to go with Michael to *Baggins, Take 5, Spång, Angel,* and *Laxtjärn*. I know that I have signed up for the longest and most exhausting route because Michael moves fast and only takes short breaks to maximise his efficiency. To update his mental map between one landmark and the next,[33] we reorient ourselves, as he usually knows the route by heart.

Today, to my surprise, Michael uses the GPS more than I have ever seen him use it. I asked him why this was the case and he answered that he wanted to get as close to the centres of the territories as possible so that Camille and Marine will be able to find the exact spot where he put the fat up when they return to this location in the following days. When guiding our way with the GPS, Michael's wayfinding expertise suddenly changes. Our route becomes less direct, and he stops several times to study the GPS; we even ski past our destination by a few metres and have to turn around.

During the rest of our tour through *Reivo*, once Michael abandons the GPS, he not only knows the most direct links between territories but also which route is best with the skis, and which shortcuts connect the individual territories. He knows which track is the most direct, what snow is easiest to ski on, where it is too steep to walk up on skis, and where zigzag skiing is required. In these moments, he changes his skiing method and, rather than drawing a straight line in the snow, he starts making sharp kick turns every few metres. The same thing occurs when we

33
Istomin and Dwyer, 'Finding the Way', 30.

go downhill. He does not seem to take new routes every year, but rather remembers the previous year's routes and follows them.

Julian, whom I accompanied in 2015, appears to be using the territory anecdotes to navigate to the territories. During the fieldwork, Michael discussed the research in general, such as the methods and objectives, while Julian would share anecdotes about his experiences over the past few years, and those of others, such as seeing an elk, which resulted in a territory called *Fat Moose* in *Fat Road*. Julian had worked in this study area before our study in 2015 and thus knew many of the anecdotes, which he could use, along with the maps in his notebook and mental maps, to guide to the territories, only occasionally relying on the GPS. When accompanying him, it was obvious that he had been in the area before and had developed a first mental map. He always knew where to go.

To me it seemed as though Julian was using the anecdotes as immaterial landmarks or vistas to remember the routes and landscapes. These stories enable him to relate to the environment, similar to how I remember *Norna* because of its difficulty, or the body of water in *Managed* that Marine and I did not dare cross. Thus, a network of narratives allows certain aspects of the field to be more easily recalled. This does not seem to facilitate wayfinding alone but also helps the biologists keep track of the characteristics of the individual bird groups. *Troll,* for example, refers to a bird that was difficult to find for several years and, as Michael told me, somehow represents all the birds living in that territory. Thus, these territory names play a much greater role than the biologists appear to give them credit for, even though they disappear when the research is published.

Maps

The maps that Michael had put up on a corkboard in the kitchen are designed in a simplified manner to represent the most essential aspects of the landscape, which allow the biologists to plan, structure, and discuss the fieldwork. They narrow down the scale of the study areas so that the biologists can conceptualise them and maintain an overview.[34] Thus, they extend the field to include the kitchen of the field house, allowing them to continue working with reference to the field even there.

All of the bird territories are marked on these maps, and Michael will mark in them the steps that have been completed each day. In the mornings and evenings, he updates everyone on the latest progress. In this manner, the maps will become the datasheets that document the progress of the research tasks. Thus, they serve as a reference to discuss the territories, routes, and tasks that the biologists must still complete. In the mornings before we leave, and occasionally in the evenings as well, Michael uses them to discuss who will go where and what to prioritise. He also uses the maps as a visual tool to provide more general information on the field. For example, they show where to park or which territories to combine when the fieldworkers are alone in the field once Michael and I have left. But for now, we remain in the field and focus on the data-collection practices that we will soon prepare for.

Following Previous Tracks

As the maps on the corkboard reveal after a few days, we have been to most of the territories in *Managed* and *Reivo* this season at least once. In the following days, we return to the territories and, accordingly, our wayfinding practices

34 Cf. Merz and Hinterwaldner, 'Neue Bilder, Modelle und Simulationen: Zwischen Repräsentativität und Produktivität', 303.

change. Today, when I accompany Marine, we do not rely on the GPS but rather follow tracks that we or our colleagues have drawn in the snow on previous days. We follow Camille's tracks, in which I can also see the traces of where she dragged her left pole when she was probably consulting the GPS. For now, we return to the bird territories for further observations. Proper data collection starts now because the first days were mainly practice to gain an overview of the study sites. The first territory Marine and I go to today is one that she has already visited. Thus, she knows where to go, and we can simply follow her previous tracks. Not having to rely on the GPS so much allows her to focus on other aspects.

Marine is much more stable on the skis now, and seems more confident and regimented in her movements and fieldwork. There is now also time to study the surrounding trees and environment, which will eventually add to her spatial cognition and implicit knowledge of the field. She comments on the sounds we hear when we stop moving, and we talk more to each other than we did on the previous days. Despite coming back to *Mader,* we do not encounter any birds today, and we continue to the next territories by following Camille's tracks to *Basilika* and *Guorbavagge.* If we have sufficient time today, we will also go to *Fello.* Marine occasionally checks the GPS to ensure that we are following the correct tracks and again on arrival to confirm that we are at the centre of the territory. However, most of the time, we encounter a piece of fat on a tree or the ground that has been left over from the day before. When we get to *Basilika,* we find birds from the wrong territory and move on, after she has identified and observed them for a bit. During this observation, Marine comments, while studying a bird sitting close to her on a branch above her head: 'It's interesting how sometimes we turn into their subjects of study and they start observing us. I wonder what they are thinking.' Once we arrive at the correct location in *Guorbavagge,* we use our skis as a bench and sit between the trees.

While we wait for the birds to arrive, Marine comments on the tools and media that are often used in biological research: 'You miss out on so much if you spend all the time looking at a screen. You see so much more and become more confident when guiding according to the environment'. After a short break, she adds, 'It's so much better to just be there and watch the nature, you see so many more things and that is how questions come up. Digital tools are often only employed because people are obsessed with the technique'. This confirms what I had been wondering all along, if there is a *Type A* biologist, who is more technology and data driven, and a *Type B* biologist, driven by the actual fieldwork and data collection, which seems to be the case with Michael's project. From this perspective, the choice of tools, media, and methods is not only based on facts but also becomes a matter of taste, preference, and worldview.

On the final day of my fieldwork, the temperature display in the car indicates −22°C after a few minutes' drive, and I take photos of the display to 'prove' it. We drive through this cold morning towards *Reivo*. Every time the temperature drops even further, I take a new photo, and I end up with several. While everyone focuses on the display by peering around Michael's shoulders, he continues driving us towards our destination. The temperature has now dropped to −27°C.

Before we separate to make our way towards the bird territories we are planning to work in today, Michael gives Camille, who I am to accompany, and me some additional advice and guidance on how to reach the first territory. On our way to it, having followed our tracks from the day before for about 40 minutes and having crashed at least once each (as the snow has become very slippery and the wooden skis are difficult to control with their telemark bindings and no edges), Camille stops at some point. She peers into the forest, turns left and comments, 'Let's try here'. To me, there was no sign of why we should 'try here'. It turns out that she had chosen the correct location to start walking up in a zigzag towards our first territory of the day, at which we arrive a few minutes later.

Now her movements on the skis and through the snow are regimented and confident. She appears to be confident in her navigation, knowing which routes are best to take and where to make a detour to get to our destination as safely as possible. Every time we get close to a territory, Camille takes out her GPS to double-check that we are on the right track. If we are close enough, and if another biologist has been there recently, she starts whistling to attract the birds. Marine told me the day before that 'now that there is fat in all territories we can whistle before we arrive. Without the food already installed, we would run the risk of teaching the birds not to listen to the whistling anymore if they then do not find any fat'. This is a new insight for me.

In the final few metres, Camille usually follows her gut, memory, or previous tracks, to find the fat that either she or someone else from the group has attached to the tree in the previous days. Occasionally, it has disappeared overnight, as ravens are also very fond of it. During fieldwork, Camille, who is usually quiet and focused when working, tells me that wayfinding is an easy task for Michael because he knows the routes by heart and is not distracted by skiing, which makes him very efficient. Camille usually follows the GPS and tracks from the previous day, and she already partly knows the territory as she has been to some places before, which helps her to navigate.

However, guiding through the field simply by following tracks does not always work. After a few days, there are so many tracks that it becomes difficult to know which ones are the correct ones. Of course, there are also tracks that are the result of someone having taken a wrong route or a detour. Accordingly, relying only on the tracks is not a solution.

Wayfinding always requires a combination of techniques that shift with the progress of the fieldwork and reflect the situated enskillment of the biologists.

Conceptualising Wayfinding Practices

Many anthropology scholars have studied the concept of wayfinding, including Alfred Gell, who differentiates between 'mental maps' and 'practical mastery';[35] Istomin and Dwyer[36] who distinguish between 'routes' based on 'vistas and transitions' and 'survey knowledge' based on mental maps; and Ingold, who focused on the 'dwelling perspective' of landscape as 'an enduring record of [...] the lives and works of past generations'.[37] However, this ethnographic study may offer new insights into the interactions between the skill of wayfinding and the tools that support this 'spatial orientation',[38] thus uncovering hidden agencies.

The process of wayfinding, as I have attempted to illustrate, is an important and highly complex requirement. However, it is neither formalised nor reflected on by the biologists. It is employed implicitly, with knowledge passed on through oral explanations and direct peer learning in the field. Wayfinding is clearly treated as a side effect. However, I argue that it influences fieldwork considerably. In this environment, the biologists cannot rely on smartphones or GPS alone to navigate between the territories. Wayfinding is also a highly sensory practice; it requires orientation skills through which survey knowledge of the study site is collected. It occurs through a combination of tools, landmarks, experience, research progress, general

35 Alfred Gell, 'How to Read a Map: Remarks on the Practical Logic of Navigation', *Man* 20, no. 2 (1985): 271–86.

36 Istomin and Dwyer, 'Finding the Way'.

37 Ingold, 'The Temporality of the Landscape', 152.

38 Istomin and Dwyer, 'Finding the Way'.

orientation skills, and skills in handling navigation tools, such as GPS and printed and, occasionally, hand-drawn maps (Figure 33).

Detours and iterations, becoming lost, and taking an indirect route are all part of wayfinding, and reflect the reality of fieldwork, which is messy, often redundant, and not always clear. Hence, wayfinding is a complex task often not captured in research papers. During the first days of research, the biologists must become attuned to the environment, tools for guidance, and practices in the field. As I observed with Camille and Marine, wayfinding turned out to be more challenging this time; particularly for the novices who were not yet familiar with the research and had to learn several practices in addition to developing a mental map and skills for orientation.

During fieldwork, I observed that new participants follow the GPS and maps closely to find the bird territories, whereas more experienced researchers can use their knowledge from previous studies to direct their routes. This changes their attention and perception of the landscape. More experienced biologists find their bearings largely through the help of natural markers, using treelines, rivers, rock formations, or other indicators as reference points. They continuously study changes in the landscape, collecting more informal information about the geographic conditions of the research territories, which then enhances their literacy of the field. Navigating between landmarks by means of maps and memory requires the biologists to study the landscape and surrounds. Landmarks can be either concrete, such as houses, paths, stones, rivers, trees, and anthropogenic infrastructure, or abstract, such as anecdotes. The better one knows a study site, the more embodied and implicit this process becomes, and the less wayfinding distracts from the data collection, birds, and features of the landscape. The GPS requires staying focused on the screen with one's head down until the waypoints have been reached.

However, landmarks and survey knowledge work only if the physical landmarks, or the context of the abstract ones, remain. Owing to certain events, such as forest management, even Michael, who has a strong mental map of the study sites, reacts with bewilderment when the forest has disappeared. Consequently, his route descriptions to others also become unreliable, especially if he does not yet know about the changes in the environment. The same can happen with a GPS. We encountered moments where the GPS indicated a thick forest, but we were, in fact, standing in a barren field covered in snow. Nevertheless, in this case, we could still use the GPS to navigate to our destination.

Besides landscape markers, the birds also help us to find the path in some instances, as we hear them twittering or moving in the trees, which usually happens before seeing them fly above us on our arrival. Accordingly, to navigate through the field, experienced biologists such as Michael and Julian use a mental map that allows sophisticated wayfinding that has developed in their minds over the years, along with a strong 'survey knowledge' that enables navigation between territories.

Depending on who I accompany, these tools and the way they are used differ. The choice of tool results in different ways of knowing the field and different metadata, as they evaluate the environment differently. The way the biologists collect route information is entirely different based on the method they use. This might not directly affect the research, affecting only the ability to get to know the field. However, it is one example of – often unintentional – filtering that occurs during fieldwork, ultimately leading to differently informed biologists. Lastly, it is highly dependent on the space and environment through which the biologists move, thus also adding to the situated practices and enskillment that constitute the capacity to do fieldwork.

4.3.2. Registering Birds

Having just drawn our last tracks in the snow before emerging at the road to which Michael had directed us, Camille and I are tired and cold. I am looking forward to going home, but Michael shatters this wish. He parks the car in front of us, greets us excitedly, and says, 'There are unringed birds in *Glottje*', and we shall have to go and catch them. He believes that this might be a new territory; thus, the field day is not over. From previous experience, I know that catching and ringing birds could take hours. We quickly put our equipment and rucksacks in the car, this time not being too concerned about removing the snow and ice from our skis or carefully putting them in the car. Time is more important right now. A few minutes after his arrival, we are already driving off towards *Glottje,* with Michael's usual, slightly rough, driving style, which does not always make me feel safe on the icy roads.

A few minutes later, he stops the car. Michael and Camille remove their skis while he gives me the car keys and tells me to pick up Marine. With the keys in one hand, I am standing in the forest next to the car that has been carelessly parked in a pile of snow, while Camille and Michael move off on their skis without telling me where to find Marine. Michael's mind is too occupied with the unringed birds. On my request, he gives me general directions on where to find Marine, and while we both have phones that would allow me to call her, I do not have a map or a smartphone with navigation with me, so I must rely on Michael's directions. In this moment I notice how dependent my daily life has become on my smart phone and digital technology.

Michael and Camille disappear into the forest, and I drive off. Just as I go past the location where I should have picked up Marine, I see her emerge from the forest in my rear-view mirror. Eventually sitting in the car together, we are relieved that we found each other. Back in *Glottje,* I park the car, and Marine and I put on our skis and follow Michael and Camille's tracks; we find them a few metres into the forest. Ringing the birds, thereby registering them for the study and making them part of the research, will be the final task for the day before we make our way home, which we will be doing in roughly two hours.

Here, in *Glottje,* Marine and Camille will learn how to ring the birds, a necessary task for consistency in the data. To note observations on the birds, the biologists must identify them as individuals; therefore, they put small, coloured plastic rings and one aluminium ring with a number code on their legs, for which Michael has permission from the Swedish Museum of Natural History in Stockholm. Most fieldworkers have experience in this ringing task because it is a widespread practice in behavioural ecology. However, ringing differs greatly according to the species of birds and their size and agility. It is something that cannot be taught in universities but has to be practised and experienced in situ (in the field). Accordingly, this next situated practice requires sensory attunement and alignment with the birds. The rings cannot be forced onto the birds, and it requires several steps – catching them being one of the most difficult ones.

Michael and Camille are already setting up the net with which they are going to catch the birds. I witnessed the catching process several times in 2015, and I stand some distance away to observe the event. Watching them, it becomes clear that, this time, it is not a collaboration between one expert and three novices, but rather three experienced field biologists. However, Marine and Camille, who are experienced field assistants, have never worked with Siberian jays, which are smaller and 'a bit fiddly', as Marine comments later once they catch them.

The thin black net that Camille and Michael have set up is fixed on two poles that divide the forest into two parts. I never measured the net, but from memory I estimate that it is 9 × 2.5 m. Thus, a 30-m^2 trap has now been positioned in the centre of the bird territory. The hierarchy between birds and biologists that is always implicitly at stake during this study now becomes manifest. It is only a matter of time before the first bird will be caught in the net; its only way out is with the biologists' help. Indeed, one bird soon becomes too excited about the food that has been placed around the net, and even in the net, as bait, and it becomes distracted and flies straight into it (Figure 34). Now its life depends on Michael, Camille, and Marine because it will freeze to death if it remains trapped for too long. Fortunately, they have mercy on the small, hopeless creature. Its death would also mean the death of data. It is in the biologists' own interest to save the jay and carefully disentangle it from the net while they wait for the other birds to get caught. The net almost literally forms a boundary between the Siberian jays as *free* living birds in the forest and the birds as research objects. Once the rings have been attached to their legs, they become naturecultures, permanently transcending the boundary between free living birds in *nature* to birds as research objects for scientific interests, as *culture*.

Catching Birds: Skilled Touch

Watching the birds fly into the nets, one can tell how surprised they are. One can almost see the fear in their eyes as their movements suddenly become more frantic. There is an exhausted flapping of wings, while the yarn wraps tighter around their small legs as they dangle upside down. The more the birds move, the more entangled they become. However, Julian had told me that the net is designed to prevent the birds becoming too entangled and getting hurt.

At some point, the birds appear to give up, drop their small bodies entirely into the net, and end up dangling upside down or lying on their backs. However, a few moments later, just when I think that they have calmed down, they attempt to flap their wings again to escape from their trap. Some downy feathers fall from their plumage, drifting slowly down until they silently hit the ground. This growing exhaustion will then be followed by the touch of two hands that, other than what they may think, will ultimately help free them; however, not before further treatments. Not only are the birds and I stressed about the situation, the three biologists also seem more alert, focused, and slightly nervous during the scenario.

Michael detangles the first bird from the net with dexterity and experience, giving it a quick stroke, and then putting one hand around its body and wings so that it cannot move much; at the same time, it seems as though his hand is also protecting the bird in an effort to calm it down. During this process, Michael often speaks calming words to the birds while he holds them up at eye level or comments on their weight and sex. If it is a heavy bird, he considers it male, while the lighter ones are considered female. With one movement, he then opens the bird's tail feathers into a fan and picks one for analysis, commenting 'it will grow again'.

Marine and Camille are standing beside Michael, observing what he does, while he comments on each step and explains how he is handling the bird. He then slides the bird into a little cotton bag and attaches the feather he plucked to the bag and hangs it on a nearby tree (Figure 35). Judging from the movement of the bag, the bird appears calm, and if one did not know, it would be impossible to tell that a living animal is inside. The second and third birds fly into the net shortly after one another; Camille and Marine remove one bird each, while Michael observes their skills, making sure they are doing it correctly and, if necessary, advising them. In general, he does not intervene much and lets them practise working with the Siberian jays.

A while later, three cotton bags – in different shades of blue and green and each containing a bird – dangle from a branch beside us. The birds sporadically become more active, briefly flapping their wings, before giving up again. With this, the first step of ringing the birds has been completed; now the actual ringing and measuring can proceed, and with it, Marine and Camille's process of enskillment.

Ringing and Measuring Birds

Michael sets up the 'ringing station' by removing his skis, something we hardly ever do during fieldwork because the snow is usually too deep and the skis are necessary to disperse our body weight and ensure that we remain on top of the snow; Michael does this only to ring the birds. However, removing the skis is also pragmatic: when the two skis are stacked next to each other, they serve as a bench. Shortly after taking them off, Michael stamps a deep hole on one side of his skis and a smaller one on the other. Sometimes, he places a small rubber mat across the two skis, but today he simply puts it on the snowy ground to sit on; he calls it his 'flexible fieldwork bench'. Before he finally sits down on the mat, he makes the hole a bit deeper to perfect his design. The small hole is now behind him so that his back does not touch the snow and he does not become wet or cold. Watching Michael sit there, it strikes me that it is rare to see this agile person, who never rests, sitting down. Camille and Marine imitate Michael, positioning themselves to his right and left. They have created a group of three benches or snow seats, where they now all sit next to each other, ready to progress to the next step of situated enskillment by registering the birds (Figures 36 a–q).

Before Michael rings the first bird, he selects three pre-printed measurement cards called ringing sheets with new colour codes on them (Figure 37); they also include spaces where Michael records the necessary measurements. Before Michael decides on one ringing sheet, he makes sure that the colour code is sufficiently different from the existing ones near this territory and those that the three biologists plan to put on the next two birds. This is important so that when they observe the birds through binoculars later, it is easy to tell them apart. Michael explains to Marine and Camille that 'pink, orange, and red are difficult to tell apart, especially once they have fainted out'. He also checks that the colours are sufficiently different from that of the nearby territories, in case they attract the wrong birds in the future, which occasionally happens. The colour codes are not chosen arbitrarily but require experience for smooth data collection in the future. This is the result of prior occasions where the field biologists struggled to identify the colour codes of some birds. Knowing that this must be considered is clearly not a result of formal training but rather the result of situated learning and knowing. Once all these aspects are considered, Michael decides on the three sheets. Camille and Marine follow this process with their own field notebooks open to check on the existing ring colours. They listen quietly as they begin shivering slightly from the cold. Now the ringing starts.

I get up to select one of the three bird bags still dangling from the tree. I do this carefully to avoid disturbing the bird too much, and bring the bag over to the group. Michael slowly removes the bird from the bag (Figure 36 a), using the same careful movement with which he had put it into the bag. With an invisible grip inside the bag, it appears as though he manages to wrap his hands around the bird's body, including its wings, in such a way that its legs are between his middle and ring finger. Knowing how to grab the fragile bird bodies in the bag is a matter of embodied skill. This can be acquired only through situated training, as this sense varies depending on the bird species and the environment, which, here in Sweden, for example, numbs the fingers and necessitates working with gloves whenever possible. However, this is only a means to an end, which will no longer be of importance to the biologists when the data are being processed in their offices.

I have observed this way of holding the birds many times in the past. This bird, however, tries to resist and bites Michael's fingers. He calls out indignantly and then uses a trick that appears well-practised: he substitutes a piece of his trousers for his finger. It seems to work and the bird continues to bite the fabric.

Once he has the bird under control, Michael puts the feather that he plucked earlier on top of his ringing kit. To avoid mixing it up with the other birds' feathers, he asks Marine to write down the colour and number code that he will attach to the bird onto a piece of tape, which he sticks in the field notebook along with the feather.

However, before this, the feather remains untouched, and Michael puts the aluminium ring with the number code and the three colour-coded rings onto the bird's legs. He explains that he chooses this order, because if the bird manages to escape, the aluminium ring itself allows for identification, whereas the colour codes work only in combination. In this way, he ensures that the bird can be turned into an inscription. Once the number rings are attached to the birds' legs, they can be individually identified; even though, for now, this would still mean having to catch them to read the numbers on the aluminium ring. Once all four plastic bands have been carefully attached to the birds' tiny feet, he closes them with glue.

Next, Michael takes a blood sample, which will be analysed by a lab to confirm or dismiss his first guess on the sex of the birds. For this, he opens the bird's wing and rubs some snow on the area that he plans to draw blood from to disinfect and numb it. He takes a small cannula to carefully puncture the bird's vein (Figures 36 f–h). He then replaces the needle with a small tube to draw a few drops of blood. This is a difficult task as blood does not circulate as easily in the cold temperature, and it requires some experience. Usually, as I witnessed many other times, Michael can do all these tasks with a bird in one hand while working with the other (Figure 36 i). Marine and Camille assist him, thus learning the individual steps.

To measure the bodies of the birds (Figures 36 j–p), Michael uses a calliper and a ruler, and Camille writes down the numbers he says out loud. He comments on each step and talks Camille and Marine through what he is doing. He says not only what it is that he does but also why, what for, how, and what to avoid.

Once all measurements have been taken and the bird has been marked with the rings, just before Michael lets it fly off, he talks to the bird one last time and says something along the lines of 'well done', 'here you go, get some of your favourite dish' (Figure 36 p). He strokes it one last time from head to tail, and the bird, which has now become calm, is set free again. Each bird reacts differently to the procedure: some fly off straight away, while others remain for a few seconds before they escape. Some twitter and bite the biologists throughout the process, while others seem calm and relaxed.

However, something has changed. The birds with the colour rings have become research subjects, but only in relation to one another do the scientists and the birds 'intra-actively co-constitute'[39] one another. In this sense, they are entangled with each other. The birds form part of the study and deliver data, but only in relation to the biologists studying them. They shape what becomes bird territory and, in turn, influence how the study areas develop. In these settings, biologists and Siberian jays are 'bodies in

39 Barad, *Meeting the Universe Halfway.*

the making' that are 'never separate from their apparatus of bodily production'.[40] In this sense, the learning environment here goes beyond the mental processing of oral explanations; it is a practice-based process of enskillment and enculturation that integrates cognitive, bodily, and sensory learning. Through this enskillment, the biologists' cognition changes and they also become part of the study. This aspect is still widely rejected in traditional discourse on objectivity. However, a situated perspective on scientific knowledge production elucidates the role of the body as sensory and bodily practice. And, with the registering of the next two birds, which Marine and Camille will do, they and the birds will become further entangled in the knowledge production and part of the scientific apparatus, as it is shaped in this study.

Marine and Camille are sitting on either side of Michael, while I am standing in front of them, occasionally sitting down on my knees when I want to be at eye level with them, and then standing up again and moving around to warm up. Each assistant holds a slightly dirty green or blue cotton bag, with two bird feet sticking out of it. They use a different technique from Michael and simply put the bag over the birds' heads and wings to keep them under control. Then Michael talks them through their first ringing practice with the jays. Marine and Camille clearly have experience. Occasionally, Michael helps accelerate the process, as everyone, and possibly even the birds, starts to grow unbearably cold. Towards the end, he takes over entirely, so that everyone is released from the biting cold that the early evening hours bring. However, through this initiation, Marine and Camille have undergone sensory and bodily enskillment to register birds and, thus, advanced one step further in their situated enskillment.

After both birds have been successfully measured, and added to the study, they are set free. This process goes faster than with the first bird; these jays rushed slightly to receive

40 Ibid., 158.

some food and a comment of appreciation from the biologists, before flying off. The four of us quickly pack up and rush to the car, which will not get warm until we arrive back at the field house. Once at the house, we revisit the field day at the kitchen table, as we do every evening. The biologists empty their rucksacks and pockets and cover the table with blood samples and notebooks to check if everything has been marked correctly. Michael shows the assistants where and how to store everything and puts the blood samples in the freezer (Figures 38 a–e).

The material on the kitchen table, which creates a direct connection to the birds, is not only evidence of today's fieldwork but also marks the starting point of a transformation: three Siberian jays whose identities have shifted into those of research objects. They are not only birds of the boreal forests anymore; they have been assigned new identities as research animals. Thus, not only have their duties changed, but a specific relationship between them and the biologists has been created, one in which the birds shift between being Siberian jays as subjects and research objects that will provide the biologists with data twice a year. As subjects, with an individual identification based on the rings, the biologists engage and interact with them, as can be seen during the ringing task. However, simultaneously, this subjectification would not occur independently from the biologists' research interest, where the aim is to turn the birds into data for analysis and interpretation. I return to this tension between subject and object, and the different qualities of the relationship between the birds and biologists in Chapter 5.

siv.
1
2

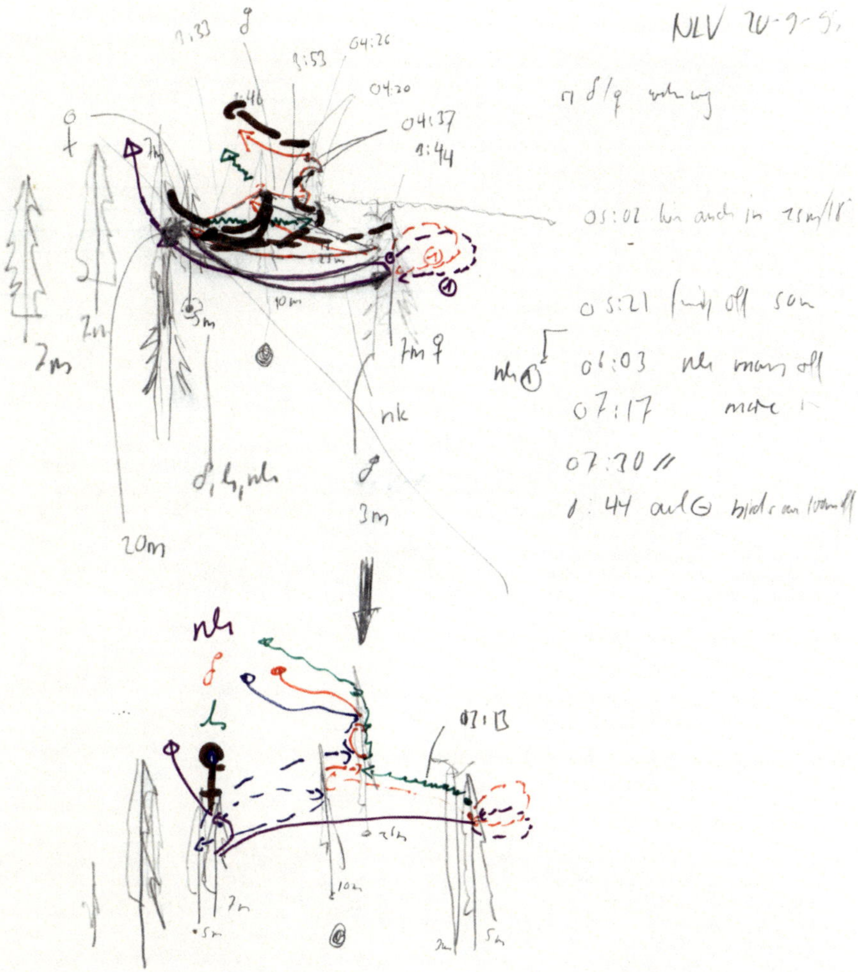
04:26
04:20
04:37
05:02
05:21
06:03
07:17
07:30
7m
2m
3m
10m
20m
nk

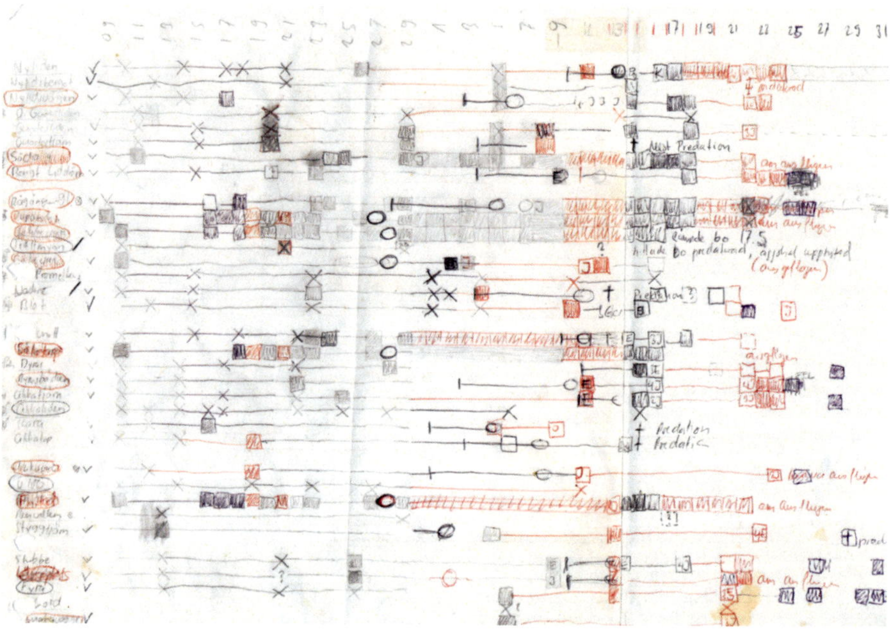

42 a, b, c, d, e, f, g, h

Figure 39:
Hand-drawn illustrations of Siberian jay movement during mating. Pen and pencil on paper. Folke Lindgren in Arvidsjaur, 1971. Zurich, 2014.

Figure 40:
Drawings of the movements of a Siberian jay group mobbing a perched hawk model, based on a video recording. Pencil and colour pens on paper. Zurich, 2014.

Figure 41:
Handwritten documentation of the breeding events in 38 territories. Colour-pen on paper. Arvidsjaur, 2001.

Figure 42 a–h:
Screenshots of videos recording the interactions of different bird groups around a feeder, as they are produced to conduct behavioural observation protocols in the office. Arvidsjaur, 2015.

4.4. Gatekeepers and Situated Ethnographers

Alongside the biologists' practices of framing and situated enskillment, I too experienced similar processes on entering the field in Sweden. As part of my participant observation, I underwent a situated enskillment when accompanying the biologists during their fieldwork. My enskillment occurred on two levels; on the one hand, I learnt the biological practices and methods of fieldwork to understand how they are employed and to follow the biologists' thinking and practices. I also learnt how to participate in their fieldwork, which forms the basis for my methodological approach. However, there is an additional aspect to this enskillment. Just as the biologists must learn to navigate the field and work with the birds in this specific study – a process I call 'situated enskillment' because, as I have illustrated, it is determined by the location in which they work, and the epistemological and ontological challenges that come with it – I, too, had to develop my own data-collection practices. These practices needed to align with the environment, my epistemological approach, and the practices of the biologists.

For this, I prepared my fieldwork by attempting, through conversations, to understand what to expect and the methods of data collection and practices of documentation to apply. For instance, based on these insights, I knew that the weather conditions – the extreme cold – would be challenging and I brought tools to combat the low temperatures and snow, which I could carry with me all day and take out of my pocket with my gloves on. However, I also reflected on how my presence, and the presence of media, might influence the biologists' behaviour.

One of the most important tools I brought with me was a small field notebook and pencil that fit into my jacket pocket, along with a smartphone equipped with an extra power bank. With the phone in my pocket, I always had quick access to a camera for photos and videos, and a voice recorder for spontaneous audio recordings. In this sense, I conducted a smartphone ethnography. I also brought a digital camera to take more focused photographs and make use of the zooming effect.

However, despite all these tools and preparations, I did not have a developed practice for dealing with the weather, moving through the field, observing the biologists, or attending to the birds and documenting my observations at the same time until that first day in the field. In this sense, during those first days of fieldwork, I found myself in the rather challenging situation of a double-bound enskillment: the biological fieldwork and my own ethnographic fieldwork. After only a few days I had developed a routine for data collection that I would refer to as a situated practice, as it responded to the conditions in the boreal forests and my research interest and, thus, my partial perspective.

While the biologists must first register their samples – the birds – I did not have to *catch* any biologists. However, I had to find out how to approach them, what information to receive from whom, and how to participate in their study in a productive way. In my case, unlike many others, access to the field was not a problem, given Michael's willingness to collaborate with researchers from other disciplines. In this sense, there were no 'gatekeepers' that I had to pass, and I was able to enter the field almost immediately. However, once I entered, I had to decide how to observe and collect data.

In the natural sciences, researchers are considered *neutral*. However, in qualitative research, the researcher is simultaneously considered an observer and the main instrument of data collection. Depending on how the researcher behaves as 'instrument for data collection', the observations that can be made and data that can be collected change.[41] My behaviour as the ethnographer can influence the roles and

41 Cf. Uwe Flick, *Qualitative Sozialforschung: Eine Einführung* (Reinbek bei Hamburg: Rowohlt Taschenbuch Verlag, 2017), 142.

tasks that the interlocutors assign me. In my case, when I was expected to train the novices during their situated enskillment, I had to be careful not to 'go native' and lose my distance to the field; moreover, it was an opportunity to understand the aspects new participants need to learn and the knowledge they may bring to the field.

As with any ethnographic study, I first had to develop a sense of my interlocutors and how they react to my presence, questions, and observations. Initially I spent much time observing and identifying their different characteristics. Thus, after a few days, I knew the best times to converse with them and ask questions, and when to simply observe. This discovery happened through implicit knowledge,[42] personal judgement of the field situation, and reading the personalities of the biologists. While I was able to make formalised observations by defining rules and following guidelines, I am also implicitly part of the research and cannot separate myself from it. As the biologists intra-actively become entangled with the birds and vice versa, I am also entangled with the biologists.

This became clear as Michael often pointed out particularly *nice* and complex datasheets and drawings in field note-books (Figures 39–41). Throughout my participation in the research project, he adjusted his data-collection methods so that the raw data would be more aesthetically pleasing. For example, he paid greater attention to the selection of frames and settings of the video recordings, thus producing not only documentation of the birds' behaviours but also *nice* videos (Figure 42). Michael had met me as a designer before I underwent training to become an anthropologist, and, initially, he recognised me mainly as a designer. He assumed that I had a particular awareness of aesthetics, thus affirming a stereotypical perception of design. Consequently, my presence in the field – as with the biologists – must be acknowledged as shaped by specific training and a particular situatedness. It greatly in-

42
Polanyi and Sen, *The Tacit Dimension.*

fluences what I can observe, and my ethnographic account cannot be reflected independently from my person. In this sense, ethnographic methodology must be understood as a situated practice that considers all agencies that contribute to the research setting and, ultimately, knowledge production.

My research practice and data are the result of an intra-active exchange between myself and the biological *other*. While the scientific result is a positivist account, the field, as I render it, is socially constructed because I attend to the social, historical, local, and pragmatic factors that influence scientific knowledge production.[43] However, although scientific knowledge production goes beyond social constructivism, it is also the result of material engagement with actors and agencies, which I aim to reveal.

4.5. Conclusions

In this chapter, I have described the practices of preparation necessary for any process of data collection in the field. Although I began by outlining the living conditions of the biologists and myself during fieldwork and our daily preparations, my focus was to assess the role of situated enskillment. I aimed to draw attention to the role of peer learning in conducting evolutionary biological data collection in the field, as opposed to formalised training in universities. My goal was to emphasise that, while university education gives biologists access to the fieldwork in the first place by providing them with basic knowledge, this training is not sufficient, and enskillment in situ is always necessary. From this perspective, fieldwork is 'down to

43
Cf. Knorr-Cetina, *Epistemic Cultures;* Latour and Woolgar, *Laboratory Life.*

earth'[44] and 'situated'[45] because it is bound to the specific field. Thus, it is not a view from above but a view from below, or rather from within.

This is the case because fieldwork differs from study to study. Regardless of what might be suggested in scientific papers, fieldwork is highly specific, given the research conditions, research objects, hypotheses, and social dynamics of a team, all of which shape biologists' practices. Consequently, to successfully collect data, the biologists had to undergo enskillment in the field. On the one hand, this enskillment occurs through practical training on the handling of tools, such as maps, equipment, binoculars, and notebooks. On the other hand, it is an embodied enskillment of the senses, and thus, also embodied enculturation.

Ultimately, biological data collection is shaped by the research conditions in the field. These conditions cannot always be controlled and are part of the research environment. Addressing them requires experience and sensory and practical knowledge, for instance, navigating the field and catching birds. The research would not be possible without this enskillment. These skills are part of a sensory knowledge seldom reflected on in the natural sciences and usually invisible over the course of knowledge production. Therefore, the research conditions and the bodily entanglement between the biologists and the research objects also disappear throughout the process. By attending to the practices of preparation, I have made the first step of biological fieldwork visible, adding one piece to the puzzle that will eventually produce the scientific results. The following chapter builds on this one, presenting another component of the research process, that of collecting.

44
Latour, *Down to Earth: Politics in the New Climatic Regime.*

45
Haraway, 'Situated Knowledges'.

Chapter 5: reCOLLECTING

The aim of the evolutionary biology study that forms the topic of this book is to collect data on the behaviour of the Siberian jays in their natural habitat in Sweden. Based on the raw data collected, new scientific knowledge can be produced through analysis, modelling, and interpretation. The biologists in this field study – Kate, Julian, Camille, and Marine – underwent a situated enskillment facilitated by their more experienced peers, primarily Michael. This involved learning how the field is structured into study areas and bird territories, how to navigate these territories, register new birds by catching, ringing, and measuring them, and how fieldwork is organised in the study. They learnt how to distribute tasks between team members and how to plan field days. They understood how to organise the data and materials, including blood samples that must be stored in the freezer, or feathers taken from the birds that must be glued into a book. Tasks also include backing up the films on hard drives and ensuring the cameras are always ready for the field days, with a fully charged battery and an SH chip with sufficient capacity. More informally, the biologists learnt how to handle their equipment during data collection. The enskillment was not only fact-based, but a bodily and sensory process of learning how things are done in this study, which I have conceptualised as situated enskillment.

In the following section, I focus on the practices of data collection themselves as they are performed in the individual bird territories. This discussion and analysis will provide insight into how sensory skills, subject knowledge, and visual framing are entangled in data collection and how they form an apparatus of knowledge production. For visual framing, I refer to the technologies that support the practices of observations, such as handwritten behavioural observation protocols and the use of cameras and binoculars.

Based on the data I collected during fieldwork, I begin by introducing the practices of approaching (5.1) that are necessary for data collection, such as attracting birds and identifying them, and setting up the behavioural observation. I focus on how the biologists observe the birds that are then finally turned into recordings, based on different modes of observation and situated mediations. Thus, I focus on the role of visual technologies that guide, frame, and support the observations, which I refer to as situated practices of an apparatus of knowledge production. Following Barad's account, apparatuses are not understood as:

> mere instruments or devices that can be deployed as neutral probes of the natural world, or determining structures of a social nature, but neither are they merely laboratory instruments of social forces that function in a performative mode. […] apparatuses are specific material reconfigurations of the world […].[1]

Practices, technologies, and bodies converge to produce knowledge. Therefore, I describe the specific observational settings (5.2) the biologists create as part of their data collection: the technologies involved, settings created, and what can be observed and how. I specifically attend to how observation tools are entangled with the practices and biologists. I focus on the practices of sensory alignment (5.3) with which the biologists engage and are imperative for working with the birds. Focusing on the senses allows me to shift the perspective from a subject–object dichot-

1 Barad, *Meeting the Universe Halfway*, 142.

omy (i.e. between the biologists and the jays) towards a sensory alignment between the birds and the biologists. While this divide suggests that the birds and biologists never engage with one another during data collection, the sensory alignment (e.g. the biologists who imitate the birds to attract them) indicates that this separation is not that clear. This means that scientific facts are not only developed by observing a *natural* event (in the case of biology), even though they usually only become visible through scientific output. This highlights the practices the biologists engage with to create settings in which they can make these observations. In these practices, the biologists do not remain *neutral* observers but must engage with activities and behaviours in such a way that they can work with the birds, for example, when attracting them. Thus, observing birds, as performed here, is less about working *on* them and more about working *with* them, namely by means of 'mimetic empathy' and adopting a 'double perspective', as Rane Willerslev[2] suggests. This is especially manifested through sensory alignment, as I discuss in the following section.

Having outlined these concepts, I continue by addressing the different human and non-human regimes and modes of observation that are present in the field. Based on the analysis of these modes of observation, I develop situated mediations (5.4) to describe the entanglement of discourse, technologies, and observations. These concepts thus bring together Grasseni's skilled visions and mediations, Haraway's situated knowledges, and Ludwik Fleck's rather historical account on thought styles and thought collectives, all of which contribute to my conceptualisation. I conclude this chapter by speculatively bringing together the biologists' behaviour observations and my ethnographic participant observation into participant behaviour obser-

2
Rane Willerslev, 'Not Animal, Not Not-Animal: Hunting, Imitation and Empathetic Knowledge among the Siberian Yukaghirs', *The Journal of the Royal Anthropological Institute* 10, no. 3 (2004): 629–52.

vations (5.5) to question the subject–object relationships created based on these two methods. This further helps address the hard–soft divide between evolutionary biology and my study.

This analysis lays the foundation for Chapter 6 where I analyse how the resulting recordings of the birds are transformed into datasets throughout the practices of fieldwork.

43 a, b, c

43 d, e, f, g, h, i

43 j, k, l, m, n, o

44

45 a, b

46 a, b, c, d

46 e, f, g, h

46 i, j

Figure 43 a–o:
Sequence of Kate identifying Siberian jays through binoculars, comparing the IDs with the field lists and territory maps in her notebook, and starting to conduct a behaviour observation protocol. Arvidsjaur, 2015.

Figure 44:
Marine during observation, looking through binoculars and holding her field notebook and pencil in one hand. Arvidsjaur, 2020.

Figure 45 a, b:
Camille and Michael observing Siberian jays through binoculars together during the first days of situated enskillment. Arvidsjaur, 2020.

Figure 46 a–j:
Sequence of Michael observing the Siberian jays with his bare eyes, extended through binoculars, field notebook, and based on video recording with the camera installed a few metres in front of him. Arvidsjaur, 2015.

5.1. Approaching

After successfully navigating towards the bird territories, the biologists manufacture a scenario to achieve their research objectives: They motivate all birds that live in the territory to come to where they have set up the equipment and feeder. Even though the practices differ slightly depending on which biologist I accompany, the overall routine for attracting the birds and setting up these observations can be described as follows.

After skiing in somewhat direct lines for approximately 20 minutes, occasionally taking a sharp turn, we arrive at the bird territories. In these moments, I can often hear my heartbeat, as the thick layer of snow absorbs all other sounds. Usually, this silence does not last long, as it is interrupted by Michael calling for the birds. With his index and middle finger pressed against his lips, he sucks air through his mouth, producing a sound like the high-pitched twittering noises of the birds' offspring. He mixes these calls with the twittering of adult birds, thereby offering a variety of calls. He simultaneously moves through the territory, making his presence known to the birds.

Then, everything is silent once more. Looking at the sky, at times through binoculars, I wait for a few seconds until the silence is broken by more calls for the birds, to which I occasionally contribute. However, imitating Michael's calls requires practice and attentive listening to the birds. I usually stop after a few attempts because I feel as though I am disrupting his calls rather than helping.

Each of the biologists' calling methods are different. Camille becomes so silent after calling the birds that, if I am not looking directly at her, I hardly notice her presence; Julian's approach is rather noisy and lively, and he frequently moves around, but he does not seem to imitate the birds nor hide; and Marine, Kate, and I occasionally use our SOS whistle on our backpacks. Despite these different methods, at some point, each of the biologists manages to attract the birds and, in contrast with what I had thought at the outset, the birds are not repelled by their presence; instead, they appear to be quite curious about the biologists.

Camille and I stand a few metres apart in the picturesque, monochrome winter landscape, and I observe her as she attempts to attract the birds. After a few minutes, the sounds of the forest begin to change. We hear wings flapping and notice sudden movements at the tops of the trees and small, dark shadows flying close to the white ground. What had been a calm and silent scene instantly transforms into a space brimming with life. Birds are flying back and forth between the fat and the tops of the trees, rushing to the ground to pick up a piece of food they just dropped, and then flying off again, only to return a few seconds later. Some appear to bump into one another while chirping excitedly. While this event does not have a pattern that I can recognise, Michael will often comment on it eagerly and occasionally respond to the jays' calls.

Once the birds are present, I become more active too. This interaction between the biologists and the birds provides valuable data, and I start documenting it. Not only do the birds fill the space with life before my eyes, so do the biologists. Thus, my research setting and observations become denser. Once Michael, Camille, Marine, Julian, or Kate, depending on who I accompany for the day, start observing the birds, I move away so that I can observe the events occurring in the boreal forests of Arvidsjaur. Occasionally, I attempt to hide behind a tree and silently observe the biologists and birds. I take out my camera, pencil, and notebook, and start observing and documenting the aspects that seem interesting to me, such as how data are collected and how the observational practices of the biologists are performed. I document their ways of studying the birds: their routines, gestures, and movements.

When I accompany Marine on one of the first days of the study, while we are standing in what I consider to be one of the most beautiful study areas – a wild valley next to a large, frozen river in *Reivo* – during a brief pause between calling the birds and waiting for them, she comments, 'You know, the birds do not appear because they think that the imitated calls are actually other birds, but because they know that it means food. They got used to it. They are intelligent birds'. She adds that she is, therefore, not entirely certain if it matters *how* one makes the calls, but that Michael is definitely more successful at it. Whether this is attributable to his particularly loud calls, his experience and implicit knowledge of how to attract the birds, or because the birds know him, she is also not certain.

In spring, after a long, cold winter, and just before mating season, the birds welcome additional food. According to Michael, they are usually desperate for it. If, despite that, the birds do not come, he believes it is because they are too far away and cannot hear us. Alternatively, depending on the territory, he speculates that some birds may be more shy or scared than others, or that they may not have noticed that the biologists have returned. This often applies to younger groups and birds that are not yet familiar with them.

The birds are more used to people in *Managed* and *Fat Road* than in the natural, protected area of *Reivo*. For this reason, Michael is always prepared to attract birds, or rather, in his words, is 'highly opportunistic' when it comes to 'getting birds'. He usually has some sausage in his pockets, which is easier to rip into small pieces than the feeder, and he believes this is the birds' 'favourite dish'.

Occasionally, Michael is also happy to attract birds that pass by when we are moving through the study site, to briefly get a glimpse of whether they are ringed or not, and, if so, identify their ID colour combinations or make sure that someone from the team will return to register them in the next few days. In this case, he writes down approximately where he saw them so that he, or another team member, could return another day to catch and ring them, thus turning them into sample birds for his study. This was the case in *Glottje* at the beginning of the 2020 season with Marine and Camille (cf. Chapter 4: 4.3.2).

The sausage is also used if birds appear unexpectedly or before anything has been set up. In this case, it works as a kind of placeholder for the fat, which is the actual feeder. In these situations, with a strong and determined gesture, Michael starts throwing the sausage into the air to maintain the birds' interest and make sure they remember that the biologists' calls mean food, as Marine had explained.

During the initial days of observation, it is particularly important that the biologists call the birds at the right time: not too early, so that they do not appear in the wrong location when we are still walking towards the centre of their territory, but also not too late, so that we need to wait a long time for them to appear. This seems to be a fine balance that differs with each biologist I accompany.

This moment of the birds' arrival is often a relief, because it means that waiting in the cold was worth it. In contrast, it can also be frustrating for me because it means that we will stay for at least 30 minutes longer to conduct the behavioural observations. There were many occasions when I was standing in the snow, unable to feel my feet and breathing into my gloves to keep my hands warm. In these instances, I gave up on more birds appearing and often even hoped that they would not appear so that we could return to the car and end the field day sooner. Days such as these often result in less detailed entries in my notebook because it was too cold to remove my notebook, take off my gloves, and start writing – or even worse, take photographs. However, the biologists, particularly Michael, are usually not bothered by the cold and continue calling the birds until they arrive.

The biologists seldom appear to give up so that they can move on and warm up by skiing again. When the weather is overcast, or later in the day, it can be difficult to attract birds, and the process becomes less predictable. Then, the protocol of waiting for approximately 15 minutes for the birds to appear, which no one seems to adhere to anyway, is suspended; and, occasionally, we wait in the cold for almost an hour, using as many tools as we can to call the birds. Sometimes this is successful and other times it is not. In these moments, the biologists' determination to collect data is the main priority, and it becomes clear that they are willing to do almost anything to achieve this.

On the first days of the season, the birds can be hard to attract when they are not yet used to our presence and have not realised that biologists with food are around. This data collection in the field typically involves several stages, where the biologists return to the territories at least twice per season. Thus, the first stage is to visit all territories once to mark their presence and prepare the birds for the following weeks. This means piquing the birds' interest by putting up fat, even if no birds appeared after calling for them, so that 'the birds know we have been here'. Although behavioural observation protocols may not be possible in a particular territory, the biologists have at least left behind a trace that the birds will eventually encounter, possibly piquing their interest and increasing the likelihood of them appearing when the biologists return. Indeed, this approach is often successful, and when we return, the birds appear more quickly; occasionally, they even seem to be waiting for us.

Identifying

As soon as the birds start arriving, they must be identified by their IDs. Identification is important not only to keep track of the birds and collect data consistently but also to ensure that the correct birds appear; in other words, to verify that these are not birds from nearby territories. This occurs occasionally and usually causes a fight, as described earlier (cf. p. 124), leaving the biologists to abort their behavioural observation protocol and return another day, or wait until the birds have calmed down.

For the identification, the biologists shift their attention to examining the birds' IDs through their binoculars (Figure 43). Michael grabs his binoculars, which hang around his neck and across his chest on a strap that resembles that of a purse. One arm goes through the loop of the strap, after which he takes the binoculars into both hands and studies the birds. An Excel sheet lists all the birds from the previous season and is glued into the back of his field notebook, helping him to know which colour rings to expect.

Often, after briefly checking the Excel list, Michael seems to remember the IDs by heart, whereas less experienced biologists take longer to identify the birds. Once he can confirm that he has seen a particular bird, he checks off the relevant ID on the list. This means that the bird has survived the previous season and continues providing data to the dataset. The most basic step of data collection is checking which birds are still there and in which territories they live; this helps to provide an overview of the study, particularly at the start of the field season.

The biologists switch between looking through the binoculars, checking their notebooks, and squinting to see better in the bright light. Their main goal is to get a glimpse of the IDs. The Excel list indicates the colour and numerical code of the rings, which are arranged according to the birds' individual territories. The order of the territories in this list is based on their geographical locations and how the biologists would usually move from one territory to the next. Territories that are geographically close tend to appear consecutively on the list. In addition to this list, the biologists have a small map glued into their notebooks on which the individual bird territories are marked with boxes where the names of the territories and the individual birds appear. While the Excel sheets are digital, the maps are both handwritten and digital because the birds' information is usually completed manually.

The biologists continue this process until they have seen all four rings of each bird (Figure 44). However, because the birds have more feathers in winter, it can be difficult to see the rings, especially when they are sitting on tree branches and their legs are covered. Michael, who in general talks and interacts most with the birds, would then encourage them by saying: 'Come on', 'Show your legs', followed by 'Not friendly' or 'Unhappy' as an interpretation of the birds' twittering. He sometimes addresses the birds with direct questions, such as 'Who are you?', while Marine usually expresses a deep sigh if she cannot see the colour IDs, and Camille calmly whistles to the birds.

Occasionally, the birds lose a ring or the colours of their rings have faded in the sun, making the process of identification more difficult. In this case, it helps to know how the colours change over time. Experienced biologists are aware that red-scale colours can be particularly difficult to work with, as pink, red, and orange become more difficult to distinguish as the rings age. If a ring is missing, they mark this on their Excel sheet with a hashtag and attempt to catch the bird again to fix it, if there is time.

Young biologists can take a long time to discern the IDs if they are not yet used to the process. During the first few days, whenever Michael and the assistants are together, they stand close to one another and look through their binoculars (Figure 45). Once the birds arrive, they start reading their colour codes out loud. All three biologists, Michael, Marine, and Camille, then confirm the colours they have identified, and, if necessary, they correct one another. When the assistants are uncertain if a ring is orange or faded red, Michael usually confirms the correct colour and comments on why it is difficult to discern and how the colours tend to change. Occasionally, though, he too struggles and must guess. Through this process, the novices exercise their visual skills to identify the colours correctly and identify the birds. This process also helps them get used to the overall referential system of the Excel sheet, territory names, and the rings, as well as working in the cold and snowy conditions of Arvidsjaur.

All the biologists I accompany during fieldwork appear to pay close attention to precise identification. They are aware that, without this, the data they collect is worthless because it cannot be related to specific birds and thus merged with the overall data plot that forms the basis for further analysis. Often, when I was satisfied with my observations, they would return and double-check that the colours were correctly identified. It is imperative that the biologists know which birds interacted and how, and not merely the fact that two birds interacted. Only through codifying their research object by means of the IDs can the biologists extract and document the information they need to conduct their research. The life data of the birds, such as their age, sex, family relationship, and status are essential to draw conclusions about their social behaviour.

Setting Up

While attracting and identifying the birds, the biologists also install the feeder, and as the birds eat, they reveal their social dynamics to the biologists. For this, the biologists do not select any tree but rather a *good* tree: one that is not too exposed so that the birds fear an attack by a predator, such as a raven or an owl, but also sufficiently open so that it is easy to observe them with no bushes or trees impairing the view. Ideally, the tree does not have too many branches to obscure the view, but enough so that several birds can perch on them and eat.

When Michael puts up the bait, he moves to the tree or bush that he has chosen, drawing a straight line in the snow with his skis. Occasionally, he rips off some branches to create a good feeding situation for the birds and the observations he plans to make, and then attaches the feeder with a wire onto his chosen spot. The team started using wire in 2019. Before that time, they would impale the fat on a spiky branch of a tree and, as a result, the fat often fell off during the behavioural observation. If Michael (or whoever I accompany) expects more than three birds to arrive (because that is what is indicated in the Excel list or recalled), he puts up two pieces so that all the birds can eat at the same time.

I noticed few branches for the birds to sit on the chosen tree, as I accompanied Marine one day. It is difficult to discern whether the jays dismiss one another, or if they simply fly away because they cannot find a place to land. Subsequently, the observations are ambiguous, and the data are not robust, as they were collected in a situation where Marine had to consistently consider additional reasons for the birds' behaviour. This observation illustrates that it matters how the experimental setting is curated. To collect valuable, robust data, the feeders must be installed in a particular way, something the biologists must learn through practice, as it is not formally taught. This is the case for the entire set-up for the behavioural observation. As far as I have observed, this information is not written down anywhere but passed on through peer-to-peer learning during collective fieldwork, which only occurs during the first few days. More likely, this knowledge is gained through trial, error, and readjustment, once the biologists notice that they cannot see the birds or that the birds are struggling to perch and eat, as was the case when I accompanied Marine.

During my second field season, Marine occasionally asked for my opinion because she assumed that I have more experience with these details, as I had joined the study before her. In other cases, when the biologists were working alone without anyone to ask, they would mention the problem in the evenings or during the drive home after a field day. Lastly, setting up depends greatly on the individual biologist's judgement of what is best for the quality of the observations and videos; for this, it is necessary to know how the data will be processed and what will be analysed. During her first behavioural observation, Marine shared that she struggled to understand what exactly to describe in the protocol and what constituted certain behaviour. She continued that she found it difficult to collect data without knowing what will be analysed.

During these set-ups, I assist in different ways depending on the biologist I am with. I usually let Michael do most of the work because he knows best and is nimble with his material, and thus the process is faster if I do not intervene. In some ways, I almost do not dare intervene. However, when I am working with others, my role changes and I help to attract the birds, put up the feeder, or, as with Marine, advise the novice biologists on certain aspects.

Camille and Marine seem to appreciate this division of labour, as it allows them to check their notebooks, identify which birds to expect, study the birds that are already present, and prepare for the observations. I take one or two pieces of fat, pick a good tree, and confirm with my interlocutors if they agree with my choice. I attach the feeder to the tree using wire, threading it through two holes I made in the centre this morning. It is usually quite difficult to attach the bait firmly, especially with thick gloves. For this reason, Michael wears fingerless gloves under bigger gloves to enable easier working. Camille, in contrast, wears thick gloves that provide additional heating. She seems even happier than the others if she does not have to take them off and allows me to do the job. However, at the end of a field day, my gloves, which I also only take off when I must, contain a greasy film of pork fat that becomes thicker over the course of the study.

As a final step in the behavioural observation set-up, the biologists, occasionally with my help, install a camera for additional recording of the event and then position themselves (Figure 46). They stand approximately two metres behind the camera, which is attached to a tripod in the snow in the same direction as their view. The camera positioning is done in such a way that they can still monitor the small screen of the camera and control the frame to see what is recorded, making sure the camera does not move, which occasionally happens if one leg of the tripod suddenly digs itself deeper into the snow. Michael also uses the time indication on the camera to see when the 15 minutes of observation are over, mentioning it occasionally on the video recording, especially at the end.

With his set-up, Michael creates a diagonal line through the forest, connecting the birds, video camera, himself, and me, at the end of the line behind him. Occasionally, I stand beside Michael, while at other times, I hide behind a tree or stand between the birds, the camera, and Michael to observe his activities. Once all the birds have arrived and the feeder and camera have been installed, Michael draws several lines in his notebook, creating the chart for the behavioural observation protocol. Standing behind him, I can hear the sound of his pencil on his notebook as he draws his lines with certainty and determination. He does this deftly, never entirely losing sight of the birds. His focus generally remains on the birds.

Before Michael starts his formal observation, he flips between the page at the back of the notebook where the Excel sheet is glued and the page where he will document the birds' behaviour during feeding to make sure he writes down the IDs correctly. Thus, he ensures that the data he will collect will be stable and referable, and simultaneously mobile, owing to the nature of the data carrier, the notebook; thereby creating continuity. In the columns of the behavioural observation protocol, he notes the IDs of the individual birds that are present. Each column represents one bird. Thereafter, he presses the record button on the video camera and starts his behavioural observation during 15 minutes of feeding.

12.9.16 14:50 -

48

49

Figure 47:
Screenshot of a notebook excerpt taken while browsing through a PDF file with my data. It shows a behavioural observation protocol with colour coding in territory *Impossible* and *Fyra* in 2016. Zurich, 2021.

Figure 48:
Hand-made wooden feeder to train Siberian jays for an experiment, one stuffed with food and covered by lichen. Wood, drill, fire. Arvidsjaur, 2015.

Figure 49:
Independent observer: installed video camera recording a fixed frame at which it is targeted for the next 15 minutes. Arvidsjaur, 2015.

5.2. Observing

Michael has just finished setting up in *Angel* when the birds move agitatedly around the feeder, approximately 10 m in front of us. While looking at the birds, I see the hill we just ascended to reach the territory. The trees, mostly pine and birch, but also some bushes covered in snow, have grown in a straight line towards the sky, forming a narrow angle between the steep hill and their trunks.

Angel is located at the edge of the study area of *Reivo* in the northeast. Its name suggests that we may encounter an angel-like Sámi man here, as biologist Barbara, together with Julian, did in 2014 when they conducted their fieldwork in Arvidsjaur for the first time and discovered the group on this territory. While Michael observes the birds, talking partly to himself and partly to the birds, I continue quietly observing the scenario. The birds are moving around on a lower branch surrounded by trees. A surprisingly large number of birds have appeared in this territory, and Michael must concentrate on distinguishing between them and documenting their interactions. His gaze shifts between the birds and his notebook in his right hand, the pencil in his left, and the binoculars, which are either in both hands in front of his eyes or dangling in front of his chest.

For the next 15 minutes, all Michael's attention is dedicated to the events unfolding around the feeder. With his full attention on the behaviour of birds, once he decides that a certain behaviour is relevant to the research, he documents it almost immediately with a code in the observation protocol. He only needs to look away from the birds very briefly at the notebook to ensure he writes the code that represents a specific behaviour in the correct box. During this process, he switches between the binoculars and his bare eyes to track as many birds as possible. He uses his macular vision, which is extended by binoculars, to look at the events and bait, and his peripheral vision to track what the other birds are doing. In this way, he remains alert to the various events until he writes down the next code in his field notebook, thus adding to the behavioural observation protocol. His dexterity and the speed with which he makes notes and observations remind me of the birds' movements, and I believe his rhythm of looking up and down may, in some ways, align with the rhythm of the birds' feeding.

Occasionally, Michael's gaze moves away from the feeder and follows the birds through the forest. I follow his gaze to see what he is looking at. Looking at the sky, I see bright blue; looking down, everything turns white; and looking around horizontally, everything turns into a dark green and brownish grey. The closer the objects are to me, the more details I can study, whereas the further away they are, the more I must guess. Nevertheless, I hardly ever use binoculars because they make me drowsy. I see pine needles scattered across the ground, branches and lichen that the wind has blown off, and some pinecones that a squirrel may have brought along. Occasionally, the snow tells the story of some mice that have crossed it since the last snowfall, while at other times, I see bigger tracks that I cannot identify. Gigantic, almost 1 m deep prints indicate that moose have come past. Besides the sound of Michael moving around and the calls of the birds, there is little else to hear. I take photographs of these details, which I often decide to delete again. I make notes in my field notebook to remember this situation, and I stop and observe, attempting not to distract myself too much by the tools I brought along.

Extending the imaginary line introduced earlier beyond where the birds are feeding, it leads above the hill into the cold, blue horizon in front of us. At its other end, it crosses a white, open field and enters the forest again. This open field is now behind me. It is covered in snow touched by gentle sunlight, which creates bright reflections on its calm surface. The forest line, consisting of dark green, almost black and grey colours, surrounds this bright white, open field. Soon we will cross it, following the imaginary line to enter the forest at its far end on our way to the next territory. We will continue this route by drawing a diagonal line in the fresh snow with our skis. We will connect our entry point into the open field with our exit point when we disappear into the forest. The line is a trace of our presence, connecting each territory until, in a few days, we will have returned to the study area several times; at that point, the multitude of lines makes identifying individual ones difficult.

While I make these observations during fieldwork, Michael, conversely, observes something entirely different. He observes one frame of the scene in great detail, which I observe from a distance. For now, he is not interested in the environment if it does not affect the birds' behaviour. He is focused on the birds, identifying their colour IDs and observing their behaviour. He knows that the weather today is good, but he does not study it – not the blue sky nor the shape of the clouds. Michael's tools for data collection and his mindset prevent him from becoming distracted and help him filter and document the relevant aspects and disregard the irrelevant.

Michael knows that to progress with his research, his data must be valuable, which means documenting the birds' behaviour as accurately as possible. He needs to turn the birds, his subjects, into data that can be transported through space and time, from Sweden to Switzerland, from here and now to there and then. His subjects, thus, become space–time entities that transcend their temporality and situatedness. Similarly, he aims to obtain an overview and sense of the birds, and to record their interactions, producing a behavioural observation protocol so that he can take them home as inscriptions, which is my aim as well.

My notes, photographs, and audio recordings are also data that I collect in the field to turn the biologists' work into inscriptions at home. However, our methods differ, as a natural scientist and an ethnographer operate in different spheres of epistemologies.

Behavioural Observation Protocol

The outcomes of the observations are usually field notebook entries with a behavioural observation protocol and video recording of the same event. For this, the biologists adhere to six different behavioural patterns relevant to their study on the social behaviour of kin and non-kin birds; they observe these and collect data on the birds' interactions around the feeder. The protocol is designed so that interactions are not recorded along a timeline, apart from the video recording. However, the data collected consist of a quantitative summary of the interactions during this 15-minute period, thus filtering out certain events the biologists focus on.

When studying the protocol that resulted from these observations, the biologists can identify the patterns demonstrated by the birds in the territory. In my research material folder, I open a PDF file with the protocol used in 2016 in *Impossible,* when I was not present in the field. I attempt to retrace the interactions observed by the biologist (Figure 47). When reading the protocol from left to right, the first code I find is the number '2', which indicates that the bird with the ID 'mb al mg mg'[3] displaced the bird with the ID 'mb al mg pi'. This means that while ID 'mb al mg pi' (as the biologist would call the birds during the observation) was feeding from the fat, the actor (as the biologists call the bird that was displaying a certain behaviour) landed where

3 The code represents abbreviations of the names of the relevant colours in Swedish, e.g., 'mb' stands for the Swedish *marinblå,* which means navy blue; 'al' means aluminium, which is the same in both languages, and 'mg' is *maringrön* or dark green.

the recipient (the bird that receives an action) was and thereby forced them to leave. I learnt that this is usually a behaviour that can be observed between non-kin birds. And, indeed, when comparing the IDs with the Excel list I have at hand, I see that the bird that chased away the other bird was the breeding male in this territory.

In the next column, I read '1111', indicating that the same bird that was chased away earlier had been waiting to feed less than 1 m from the feeder – denoted by the code '1'. At the same time, another bird, identified as 'li # o al', the offspring of the breeding couple and thus a 'kin' bird, was feeding. During the 15 minutes of observation, this event occurred four times. One box to the right, notes '0100', indicating an interaction between the bird that thus far has not been able to feed at all and the breeding female bird, ID 'gu al r #'. The bird joined the female while she was eating, and the female accepted this, as indicated by the '0', which means 'feeding together'. These two birds ate together three times during the 15 minutes of observation, and once, as indicated by the '1', the other bird was waiting nearby.

As this same bird was also displaced from the feeder three times by the offspring of the breeders, I conclude that it must be a non-kin bird – born in a different territory and chased away by its siblings. It appears to have later joined this family, creating what the biologists would describe as a 'typical' family dynamic involving both kin and non-kin relations. One final type of behaviour was observed in *Impossible,* indicated by the '3' in the bottom right. This code means that the breeding female was physically chased away by ID 'lg li al r' – either through bill snapping or being pursued in flight – after first being displaced by the same bird.

The chart allows the biologists to draw initial conclusions about the birds' hierarchical family structures. These will later be confirmed or dismissed by a DNA analysis and the video recording of the event. However, as the analysis may take place over several months, the preliminary observations allow for an initial conclusion. This is important as, depending on this, the biologists will select certain bird groups with which to conduct further experiments during their field season. A *good* group for experiments is, for example, one with high family activity, with many kin and non-kin birds in the group and diverse relationships.

The experiments follow the same procedure as the basic behavioural observation protocols. They usually identify whether a hypothesis – an assumed explanation for the relationship between behaviour and reason – has been formulated correctly. Accordingly, many experiments enhance situations that the birds encounter in their environment by creating artificial settings. In this way, the biologists both provoke the birds' reactions through the experiments – enabling observation and documentation – and use these experiments to identify the underlying factors that produce such behaviours.

During experiments the biologists usually manipulate physical aspects; in this case, the birds' environment. This starts by providing them with food that otherwise would not be available to them. Often, but never when I joined the study, the birds are exposed to settings where they are either confronted by a dummy of a predator or the predator's calls. Alternatively, to study how the birds learn from one another, in 2015 the biologists exposed the birds to a self-built feeding box (Figure 48) and recorded their behaviour with videos and notes. Regardless of the events observed, the biologists consistently use behavioural observation protocols to record the birds' behaviours.

The process of video recording is a combination of observing the birds visually, while making behavioural observation protocols, which they supplement with an audio recording. They verbally describe all interactions based on the codified categories, along with everything else they can observe, particularly aspects that occur outside the video frame. Thus, the events they audio record are not arbitrary but highly selective, corresponding to the codes defined in the behavioural observation protocols.

The biologists start the video recordings by stating the date, time, and territory, and naming the birds they have seen. Some add their name so that other biologists who might analyse the recordings can return or at least know who made the video. This is followed by which bird is on the feeder and in which position. Saying all four colours of the birds' IDs takes too long, particularly if events are occurring rapidly. Thus, most biologists, once they have identified the birds, continue referring to them by their most prominent ring colour, along with their sex or family relation. This results in names such as 'orange female' or 'breeding female' rather than 'light blue, orange, metal, pink' for the video recording. Before the biologists finish their observations in the field, they often make some final notes in their notebooks, disassemble the camera, and store everything in their bags.

While the biologists solely used the video recording to record the birds and thus collect data, the behavioural observation protocol is also a significant tool for training the biologists' observational skills. Employing the protocol can be a complex task involving identifying birds, noting behaviours, and documenting observations. Thus, the biologists develop their visual attention and train their eyes within this framed event. It is a tool that allows for a quick overview of the different behaviours within the individual families and to learn, understand, and interpret these behaviours. However, for this, the biologists need to learn to translate between the behaviour and the relevant code, and the birds and their IDs, to make sure they document their observations correctly.

Once Michael has finished his behavioural observation protocol, before leaving the territory, he makes sure that the feeder is attached securely to the tree and that it is 'raven proof', which means that ravens cannot attack the jays too easily when feeding from it, and that the ravens cannot simply steal the feeder. Subsequently, he and the other biologists move from one territory to the next, following the same routine to attract the birds, waiting until all are present, and then conducting their observations, while I follow them on my skis through the tracks they have created in the deep layer of soft snow. At times I can see them, while at other times I merely follow their tracks. When I hear them calling for the birds, I know that they have arrived and that the next territory is close.

From Bird to Data

This moment of documenting and codifying behaviour is a crucial turning point for the animals' transformation from birds to data. Only by documenting a specific activity, performed by the birds and observed by the biologists, does this activity become data. Before this, it is an ephemeral event entangled with many others that must first undergo the selection process by the biologists, who decide whether the observed behaviour is valid for the research. Once the behaviour has been captured as data, the birds are turned into a code and a datum for the overall dataset. At that point they have been 'tamed' from wild birds to scientific objects. Now the birds themselves as living creatures in the forest only continue existing in the memories of the biologists as testimonies of the observations they made. The more the biologists physically distance themselves from the field, the more the birds disappear from the study as *birds* for the sake of their inscriptions as codes and numbers. These small codes in the field notebooks are the starting point of an entire body of scientific analysis that unfolds based on this documentation and the biologists' decisions to note them down.

The research objects of the biological case study, the Siberian jays, are thus turned into inscriptions that represent a specific aspect of their behaviour, namely one brief sequence of interactions during feeding, which will later be interpreted by the biologist as the social behaviour among the group. All other actors, such as other animals or bird species, possible predators, the few people that may be encountered in the forest, as well as changes in the landscape and problems faced during the observations, are partly noted down as side-notes but mostly they are ignored. The side-notes serve two main purposes: First, to contextualise the observational setting, which helps the biologists to remember the field situation; and second, as metadata, particularly on predators and settings. The biologists will return to these notes if something unexpected appears during data analysis.

These side-notes usually do not become data but contribute to an overarching framework of knowing, understanding, and interpreting the field. The notebooks' capacity to record side events draws the biologists' attention to these moments. In doing so, the notebooks exert a form of agency that shapes observational methods, contributes to the field experience, and highlights aspects of research that go beyond quantitative data – revealing the extraordinary within the everyday.

Modes of Observation

From an ecofeminist perspective, behavioural observation studies are part of an apparatus of knowledge production. This apparatus is shaped by the relevant discourse, epistemologies, and ontologies. As Barad establishes, 'There is something fundamental about the nature of measurement interactions such that, given a particular measuring apparatus, certain properties become determinate, while others are specifically excluded'.[4] Accordingly, what is decisive about this apparatus is that it not only attends to specific properties but also helps to constitute them – particularly in the case of the birds' behaviour. These aspects can only be rendered as data through processes of selection, which necessarily involve the exclusion of others. Barad continues: '[w]hich properties become determinate is not governed by the desires or will of the experimenter but rather by the specificity of the experimental apparatus'.[5] What follows is that (only) based on the properties of the apparatus are the biologists able to collect their data. In doing so, they become as much a part of the apparatus as the discourse that has led to the definition of methods and research questions, thus shaping the requirements for the apparatus.

Thus, the apparatus in my case study involves the way Michael extends his vision with the binoculars. It is determined by his vision, which is framed by the behavioural observation protocol, which is again informed by the behaviours of the birds that will be documented. At the same time, the protocol determines how the behaviours of the birds are defined. Therefore, the behavioural observation protocol is more than just methods, hypotheses, biologists, and research *objects*. These are part of an apparatus in which the biologists, based on their bodily practices (of looking) and epistemological practices (of knowing what to attend to), intra-actively co-constitute one another together with the birds. In these settings '[...] neither the subjects nor the objects of knowledge practices can be taken for granted, and [...] one must inquire into the material specificities of the apparatuses that help constitute objects and subjects'.[6] To conclude Barad's argument, the birds, biologists, and experimental apparatus are inseparable, and changing one

4 Barad, *Meeting the Universe Halfway*, 19.

5 Ibid.

6 Ibid., 27.

aspect in one entity influences the entire process of knowledge production. During data collection, material-bodily practices and cognitive-epistemological practices are no longer separable. The biologists, jays, technologies of observation, and discourse together form an apparatus, and the resulting inscriptions of raw data are the phenomena created by this intra-active process. Ontologies and epistemologies of the apparatus thus become entangled into onto-epistemologies (Barad).

Against this background, I analyse the properties of the 'apparatuses of visual production'[7] that shape data collection in my case study. With regard to the behavioural observations, I suggest that three modes of material–semiotic vision can be identified, '[…] including the prosthetic technologies interfaced with [the biologists'] biological eyes and brains'.[8] First is the biologists' natural, or rather bare, vision based on their human eyesight, shaped and formed by their formal biological knowledge and their situated knowledge within this study. This vision is informed by the ongoing scientific discourse and the result of situated enskillment of the study. Second, the biologists' observations are, as skilled mediations, shaped by technologies of vision that '[…] function as catalysts of our attention and action […]'.[9] In these skilled mediations, material practices and technical skill (of handling, e.g., the binoculars and behavioural observation protocol) are combined with visual skill. Together, they form the epistemological practices of observing, categorising, and documenting the birds' behaviour. Third, in terms of the video recording, the visual task of observing is delegated to an automated device.

These three ways of studying visible events result in three kinds of vision: framing, filtering, and, eventually, data. They result in different intra-actions between the biolo-

7
Haraway, 'Situated Knowledges', 589.
8
Ibid.
9
Grasseni and Gieser, 'Introduction: Skilled Mediations', 13.

gists and technologies, different ways of storing data, and different bodily practices. While bare vision is limited by its embodied frame, natural focusing, and the inability to zoom or record, tools like the video camera and binoculars extend these capacities, enabling the biologists to see in new ways. However, the case of the biologists shows that only in their combination is data collection, as it is done here, possible. Thus, all aspects are part of the apparatus of observation.

Bare vision is not extended by any tools or media. It is, however, a mediated form of perception, shaped by the human body, biological limitations, and the visual apparatus itself. While it is the same visual tool the birds employ, they *see* differently in different colours, contrasts, shapes, and dimensions. It helps guide us through the territory and see risks before it is too late, and it allows us to keep track of what we want to follow on our skis. The biologists and I also use our bare vision to study notes, maps, charts, and data, and to take notes and fill in the behavioural observation protocols. The difference manifests itself in *what* we see in those materials as a result of our (epistemological) situatedness. Even observing and identifying the birds with the naked eye – and distinguishing certain behavioural patterns from others – requires professional training, initiation, and knowledge. It is, in this sense, the biologists' and my *natural* vision. However, it should not be *naturalised* because, returning to Haraway, it is a situated practice, a partial perspective, and a view from somewhere. It is not objective, but always part of an apparatus, and the result of training, schooling, and skill, and thus never *neutral.*

What can be observed is not arbitrary but determined by visual skill and situated enculturation. When the biologists arrive at the bird territories, they begin by scanning the environment, visually and auditorily. They study the trees and the surrounding landscape, looking and listening for birds. While I focus on the biologists, at times imitating them by also looking up – albeit with less motivation to see birds – I want to recreate their gaze to understand their methods of observation. However, whenever I attempt this, I notice a boundary where my observations stop. I often struggle to actually observe the birds, and I cannot interpret what I see from a biological perspective, or I do not even know what to focus on. Accordingly, in this field situation, despite being in the same place and exposed to the same events and practices, whether it is the biologists' visual attention or mine, we make entirely different observations, as we come from different ecologies of practice and are part of different apparatuses of observation.

While much can be observed with the bare eye, once it comes to formalised data collection and specific observational practices, its limits may be reached. The naked eye can no longer do what is required by the biologists to turn the birds into data. Bare vision is distracted too easily. It does not provide sufficient focus for the biologists' observations, and it lacks the capacity to store information. In addition, there are differently trained and enskilled eyes involved in fieldwork. To compensate for this difference and create consistency, the biologists' visual skills must be aided by technology. They must become mediated, not only by the discourse but also, in its most literal sense, by tools and media. To collect data consistently, bare vision must be extended or complemented by binoculars and the behavioural observation protocol, which help guide and narrow the focus.

The binoculars frame the biologists' vision and enhance their natural visual capacity. The main feature of binoculars is their capacity to bring things closer. This allows the biologists to overcome visual distance that would otherwise not be possible. Binoculars help the biologists to look more deeply into the forest and higher up into the trees. They can observe events on a larger scale and with richer detail than they would naturally be able to. The powerful zooming effect makes the birds appear to be right in front of the observer, thus also affecting their scale and image. This is necessary to identify the individual birds and accurately study their interactions during the behavioural observation protocol.

However, binoculars create a blind spot, a visual gap between the biologists and the birds, and thus, significantly tunnel their visual attention by excluding their immediate surroundings. In this sense, the binoculars help guide their attention by making the biologists blind to the complexity of their surroundings. However, the binoculars do not facilitate this as detached entities. They do not work separately from the biologists' bare vision but in intra-action with it. Thus, the observations are shaped by this combination of vision. In this sense, the gaze that is extended through the binoculars is the same gaze that is always also epistemologically and ontologically situated, as described earlier.

From this perspective, the binoculars must also be considered part of sensory enskillment because they require a different sensory attunement than the bare eye. Looking through binoculars must be learnt. Only once the biologists have mastered it, overcoming visual distance without physically changing their *standpoint,* have they been sensorily aligned to the capacities of their tools and are they able to *see.* Without experience, it is difficult to bring an object into focus when looking through binoculars, particularly if the object of study is as animated as the birds and moves as fast as they do. Therefore, looking through binoculars – besides the knowledge of what to look for – requires visual skill and training in the interaction between the body, eye, and tool.

Once these steps have been completed, an additional tool for tunnelling the observation can be added to the observational apparatus and the behavioural observation protocol. As described at the beginning of this chapter, the protocol determines the properties that become data by drawing primarily on the observational chart, thereby including only these aspects and effectively excluding all others. Now, the biologists observe the birds with eyes (their extended vision by means of the binoculars), and hands (the notation practices of the protocol). These practices have become inseparable and co-constitutive, functioning as situated mediations that shape the observational entanglements between the birds, as research objects, and the biologists.

The last mode of vision that accompanies the observational apparatus in the field is the video recordings. They allow the biologists to zoom in, document, and store a large volume of detailed data on the birds' behaviour. The biologists can go back and forth during analysis once they have transported their data back to their offices, given that the recordings document the entire event and are an essential part of their raw data.

The videos follow a different logic of visual attention. The biologists set up the camera in advance and the camera steadily observes the events in front of it, despite technical glitches that may occur. This documenting process occurs independently once the record button has been pressed, until the biologists intervene. From then on, every detail is documented and stored on the SH chip of the camera; thus, it is like the behavioural observation protocol but everything that is outside the frame is filtered out. While the video camera does not differentiate between relevant and irrelevant aspects within the frame, it recreates a thick documentation of the event in a linear time frame. The behavioural observation protocol, in contrast, creates a quantitative extraction of only the relevant aspects, which simultaneously become visible, thus also filtering out the aspect of time.

In addition, the video camera detaches the bodies of the biologists from the observation once the documentation process has been set up (Figure 49). Thus, visual skill is not relevant here, except in choosing the frame. However, the biologists compensate for the limitations of the camera by verbally explaining the observations, particularly those beyond the frame. They focus on naming the IDs of birds that are interacting and emphasising aspects that are relevant. These descriptions are not arbitrary, but, again, follow the logic of the behavioural observation protocol. The biologists' coded observations are subsequently verbalised and recorded as part of the audio layer. Only once these data have been transferred to computers, do the biologists return to their bare vision to analyse the recorded material in their offices, which I shall discuss in Chapter 6.

All modes of looking – the camera and the (extended) eyes of the biologists – simultaneously observe the birds. Together, these modes of observation follow a hierarchisation of visible events to determine what exactly is turned into data. As a form of multi-layered filtering, the combination of these methods allows the biologists to translate bird behaviour into quantitative datasets, which in turn form the basis for scientific analysis and results. They constitute the apparatus of observation. As a layering of visual practices, the observations become a visual description of the social interactions of the birds, which then forms the basis for further analysis and production of robust data.

The observational system is established in such a way that it allows the biologists to record, document, and store a thick description of their observations on transportable media – immutable mobiles – that they can transport from Sweden to their university offices for interpretation and analysis.[10] These visions add up to a metaphorical *seeing* in the sense that the biologists, depending on how they combine the data and results of visual practices, gain new knowledge and in*sights* within this apparatus of knowledge production. If successful, they can *see* something new in their resulting data, which will then lead to new publications.

10 Latour, 'Circulating Reference: Sampling the Soil in the Amazon Forest'.

Situated Mediations

As I have illustrated, the practices of observation in this case study require highly skilled vision, rely on prior biological knowledge and training, and are both situated and mediated. Drawing on three different concepts, two of which have been introduced in the theoretical framework of this book (Grasseni, Haraway, and Fleck), I conclude by introducing the concept of situated mediations. This concept will help explain how professional vision is determined by apprenticeship, and how visualisation technologies and biological vision co-constitute and intra-act with another. Thus, they do not replace skilled visions, but instead become a property of situated mediations. They facilitate how the context in which vision is employed defines what can be seen, thus becoming part of the knowledge that will eventually contribute to the scientific discourse and define what is disregarded and epistemologically obscured.

While Grasseni highlights vision as a professional, enskilled, and culturally determined technique, Haraway expands on this concept politically by emphasising that vision is never neutral and always situated. No matter if it is embodied vision or the technologies of vision, in either case, the visual capacity is mediated. What Grasseni calls skilled visions as a professional practice can be identified as one aspect of what Fleck called a 'thought style' that situates and defines a certain way of thinking within a thought collective, usually formed by representatives of a particular discipline:

> We can […] define thought style as [the readiness for] directed perception, with corresponding mental and objective assimilation of what has been so perceived. It is characterized by common features in the problems of interest to a thought collective, by the judgement which the thought collective considers evident, and by the methods which it applies as a means of

> cognition. The thought style may also be accompanied by a technical and literary style characteristic of the given system of knowledge.[11]

Fleck also emphasises that vision is always directed and, thus, dependent on disposition, style, atmosphere, and perception, and on defining what is evident. He emphasises how the discourse is blind to external phenomena. Or, to put it in Margulis' words: 'Any idea we conceive as fact or truth is integrated into an entire style of thought, of which we are usually unaware'. This can be understood as 'the cultural constraints', 'trained incapacities', 'thought collectives', and 'social constructions of reality'.[12]

With situated mediations I focus on how the observations of the biologists are determined not only by the technological mediations, based on binoculars, cameras, and protocols but also by discursive, institutional, and political mediations. Situated mediations are the result of a thought collective. Thus, 'they affect all of us, including scientists. All are saddled with heavy linguistic, national, regional, and generational impediments to perception'.[13] Consequently, they are also affected by the individual situatedness of the biologists. These aspects cannot be treated as separate, and scientific observations are always the result of a partial perspective on material and immaterial practices. They are a co-constitution of body and senses, and of ontologies and epistemologies. Situated in relation to the situatedness of the biologists, they are the institutional dependencies, such as funding, the situatedness in the actual field, and the knowledge, experiences, and bodies of the individual biologists. They are mediated through the thought collective, bodily practices, and the senses, but simultaneously also through actual tools and media.

11
Fleck, *Genesis and Development of a Scientific Fact,* 99.
12
Lynn Margulis, *The Symbiotic Planet. A New Look At Evolution* (New York: Phoenix, 1998), 3.
13
Ibid.

The following three aspects are characteristics of situated mediations in my study. First, situated mediations as skilled visions are characterised as the ability to differentiate between important and less important events during observation and help to filter out – for the biologists – irrelevant information. Based on these, the biologists know what to focus on and what to disregard. They know how to observe the birds based on specific practices, tools, and visual technologies. This direct vision is supplemented by a secondary peripheral vision that allows the biologists to track events occurring outside their focus and thus maintain an overview. Second, the visual realm is accompanied by other senses, such as the auditory – for example, when calling the birds – which further helps to attract and identify them.

As part of the skilled visions, the biologists employ several tools and (bodily) techniques that culminate in a multitude of visual practices to study the birds' behaviours. Thus, this vision is supported by tools and media. These serve two purposes in this case. They work as extensions of the natural sensory capacity and extend the vision of the biologists, as I have discussed, so that they can see what they need to see and would not be able to see without it. In addition, they structure, frame, and guide the visions of the biologists. They do this based on 'focusing media'[14] such as behavioural observation protocols, by limiting the space of action to a minimum in the moment of the research activity. Together, the skilled visions and notation techniques help the biologists to filter, separate, and distil the information that is relevant for their research. They determine the visual capacity and what can be observed.

Third, skilled visions are learnt; from this perspective, they are a mediated process among biologists of the same thought style. The visual capacity is determined by their prior knowledge and their tasks, interest, and experience. They are situated within the body – thus embodied – and

14
Grasseni, 'Skilled Vision: An Apprenticeship in Breeding Aesthetics', 44.

in the direct environment in which the visual observation is occurring. This process reveals 'the fact that knowing does not come from standing at a distance and representing, but rather from a direct material engagement with the world'.[15] This direct material engagement is the result of a particular process.

In conclusion, these situated mediations 'structure [and frame] the context of action'[16] during observation, making scientific insight possible. The binoculars, notebook and pencil, and video recordings thus become extensions of the biologists' bodies which allow them to translate their study object into visual inscriptions. While the data on the birds become richer to the biologists with each visual inscription, the inscriptions allow for categorisation and classification into a bigger, overarching system: a taxonomy. Simultaneously, the visualisations themselves are formalised conceptualisations that simplify information. The birds – with their emotions, thoughts, social relations, and being – are reduced to statistics, data, and charts. However, without this reductionism, no scientific insights would be possible, as the *un*translated research object is too complex for quantitative data analysis. Without these mediations, from the biologists' perspective, data collection would not be possible, nor the production of new scientific knowledge. Finally, situated mediations are a prerequisite for scientific fieldwork, as in my case study.

5.3. Sensory Alignment

Data collection in the field depends on multiple sensory alignments between humans, birds, and environments. Thus far I have mainly focused on the visual observations

15
Barad, *Meeting the Universe Halfway*, 49.
16
Grasseni, 'Skilled Visions: Toward an Ecology of Visual Inscriptions', 32.

and alignment of the biologist's gaze with the constantly moving birds. Alignment, however, also refers to the other senses involved in data-collection practices and fieldwork.

To understand how sensory alignment is employed by the biologists during fieldwork, I refer to Rane Willerslev's[17] work on 'mimetic empathy' by assuming a 'double perspective', which he developed based on his ethnography of Yukaghir hunters, thus working in a similar geographical location and environment to the biologists and me, albeit with a different epistemological focus. While he describes these concepts as techniques for hunting, I apply them to the biologists' attraction of the birds. Although scientific observation has nothing to do with hunting, the sensory practices that both hunters and biologists engage with prior to the killing – or in my case, observations – are similar. In both cases, they must attract the animals using the least invasive practices in their habitat where they, the animals, are the experts and the humans inferior.[18]

Mimetic empathy, as Willerslev describes, '[...] is the ability to put oneself imaginatively in the place of another, reproducing in one's own imagination the form of the Other's perspective'.[19] The biologists, who I do not aim to portray as hunters here, perform similar practices to the Yukaghir hunters in attracting the birds. This means that it is essential to adopt the perspective of the birds. However, the extent to which this perspective should be adopted has its boundaries; the aim should not be to adopt the birds' perspective in an absolute sense, according to Willerslev, who describes hunting as an Indigenous tradition that follows

17
Willerslev, 'Not Animal, Not Not-Animal'.

18
With this, I am not implying that the observations and experimental settings that the biologists create in the forests do not affect the birds or their results; instead, I use the term to distinguish it from laboratory conditions in which the research objects are isolated from their environment and the system in which they are usually situated. However, what does not seem to be reflected is the presence of the biologists when they observe and study the birds and how it may affect their measurements.

19
Willerslev, 'Not Animal, Not Not-Animal', 647.

a different sensory alignment to non-Indigenous relations to nature, particularly in Western thinking. Rather, the biologists must adopt a 'double perspective':

> [T]hey attempt to assume the point of view of the animal, while in some profound sense remain the same [human]. Mimetic practice, [...] provides the ability to be like, yet also different from, the animal impersonated; it grants the hunter a 'double-perspective' whereby he can assume the animal's point of view but still remain a human hunter who chases and kills the prey.[20]

For the biologists and their interests, this means that they attempt to think as the birds do in order to work with them; however, I would not assume that they attempt to impersonate them as Willerslev suggests for the hunters. They aim at technically and sensorially manipulating their environment, including themselves, in such a way that they can observe the birds. For this, they create their observational settings, which are designed so that the birds can display their usual cooperative behaviour, even though it is an artificial setting. With reference to anthropologist Michael Taussig's concept of mimesis, this sensory alignment by means of mimetic empathy, '[...] collapses such dichotomies as Self vs Other, nature vs culture, and essentialism vs constructionism'.[21] By doing so, the biologists create a multispecies *we* during fieldwork, a sensory entanglement between the birds and themselves, that shape one another, which Barad refers to as intra-active becoming.

These two concepts – mimetic empathy and double perspective – permit analysis of the different sensory alignments of the biologists with the birds during fieldwork. Other than for Willerslev's hunter, the sensory alignment of the biologists mainly occurs by means of visual, tactile, and auditory perception. Highlighting the role of sensory

20
Ibid., 630.
21
Michael Taussig, *Mimesis and Alterity: A Particular History of the Senses* (New York: Routledge, 1993), 252 as cited in Willerslev, 'Not Animal, Not Not-Animal', 639.

alignment during biological fieldwork will also help to shift the perspective of scientific knowledge production from one that is only based on objectifying the animal *Other* to one based on engaging with their *personalities,* if you will, a prerequisite for conducting fieldwork in the first place that should be conceptualised as such: the cooperative behaviour of the Siberian jays in their family groups. These personalities become visible in their interactions and when Michael would informally categorise them and comment on the individual birds' behaviours as well as that of the groups.

Attracting Birds

As described in the previous chapters, the biologists find the birds in their natural habitat. They immerse themselves in the biological world of the boreal forests. This immersion only works through sensory alignment. First, to navigate through the forests in the best way and arrive at the bird territories, the biologists need to understand the environment in which they work from the birds' perspective.

Second, as the birds in this study are not usually equipped with any radio tags that would allow the biologists to track them, they must follow them in other ways. Accordingly, the alignment in this case starts with attempting to understand the way the birds inhabit the forest; the routes they fly, the parts they avoid, what makes them feel comfortable, and finally, which territories they choose. The biologists must learn to empathise with the birds; when looking for their territories, they must think like birds, particularly when deciding where to put up the feeder, so that the birds feel comfortable.

This alignment becomes particularly visible when attracting the birds. Once the biologists arrive in their habitat, they attune themselves to the birds auditorily. They do this mainly by imitating the birds' calls, in terms of melody, rhythm, and intervals. One could claim that the biologists attempt to sound like them. Skilled biologists such as Michael imitate an entire partiture, mimicking their sounds and offering them a range of calls. Michael, with his many years of experience listening to the birds, switches between the imitation of the birds' offspring, and other 'happy' calls, which he calls 'chit chat', avoiding whistling in a way that the birds may interpret as a warning call or that of a predator. This way of communicating with the birds is only possible through mimetic empathy, deep immersion, alignment with them, and an understanding of their calls, which can only be gained through experience and extensive training, particularly when it comes to imitation. The biologists also remain alert for the calls of predators – most often ravens – during data collection. They usually do not see predators because they are focused on the jays.

In these moments of auditory alignment, Michael appears to be exhibiting what Willerslev refers to as the double perspective. He imitates the calls with a rich pattern of Siberian jay sounds, while somehow moving through the territory in a light and animated manner, like the birds themselves. His experience appears to have reached a point where he adapts his bodily actions to the research, along with adopting the behaviour of the birds, thus, literally embodying the research and incorporating it in his bodily knowledge, beyond the imitation of calls and observational practice.

Through this mimicking performance, he appears to become the bird, while at the same time maintaining focus. As soon as he notices a bird, he adjusts his calls, for example, by waiting for the bird's response, but also by grabbing his binoculars. He looks through them with the clear intention to identify his research subject by examining its IDs. He does this while continuing to call for the birds. In these moments, he performs Willerslev's double perspective: transcending the boundary between human and non-human through auditory and partly physical alignment with the birds; simultaneously performing and manifesting difference based on his visual observation as biologist, in which the birds are still a source of data.

Calling for the birds also involves knowing when to start and at what point it is necessary to feed them sausage to make them stay. This requires an alignment or empathy, not only as an auditory experience but also as the ability to understand the jays' thinking to a certain extent. This kind of understanding led Michael and other biologists to the conclusion that jays are easy to study because they are relatively tame and curious birds that are easy to attract.

In Michael's case, he may 'transcend inter-species barriers with the aid of [his] embodied imagination [...]'[22] in the moment of attracting the birds. However, this is not the case with all the biologists I have accompanied, and it appears to be a question of experience, practice, and extensive immersion in the field. Nonetheless, younger and less experienced biologists without this 'doubleness', such as Julia, Marine, Camille, and Kate, were able to successfully conduct fieldwork, occasionally even through arbitrary whistling that did not have much to do with the birds' calls. As Michael told the group, it is not so much about imitating the calls but rather revealing their human presence. Based on my other interlocutors' comments, I would still claim that Michael's success in attracting the birds, and his efficiency and speed in fieldwork, is based on this mimetic empathy and double perspective.

Thus, in varying degrees, the biologists '[...] project themselves, through practices of mimetic empathy into the life-world of [the birds]. Practices of mimetic empathy provide the entrée, so to speak, to the perspective of the animal'.[23]

22
Willerslev, 'Not Animal, Not Not-Animal', 648.
23
Ibid.

Lastly, sensory alignment is not a formalised scientific practice, but a mindset that accompanies the research, and – in both cases – the biologists may return home with data. The practices they employ still differ, and the ontological boundaries in which they are situated during fieldwork shift and overlap, depending on the degree of mimetic empathy and double perspective.

Sensory Alignment and Registering Birds

This intra-relational and intra-species 'becoming with anoter',[24] as it becomes particularly visible in the moments of sensory alignment, is ruptured when registering the birds. Then, instead of aligning with the birds by making them offerings that they can deliberately accept or reject, the biologists impose their will onto them, by aligning with them not as an offer but a trap (cf. Chapter 4: 4.3.2). They literally install a trap that the birds will fly into, briefly falling prey to the manufactured structures in their life-world. In this moment, the power that this mimicking affords the biologists becomes visible.

This sensory and emotional alignment enables the biologists to understand how the birds fly through the forest and how they think, which informs them about how to set up the net and catch the birds. And while, on the one hand, the biologists are fascinated by the intelligence of the birds and their personalities, such as being 'shy' or 'friendly', on the other hand, if the birds' intelligence helps them avoid the nets, they become frustrated. They then encourage the birds by shouting 'Come on!', or if a bird is close to the net, 'Now, now, now!', and if they do not fly into the trap, a frustrated 'Noooo...' follows. These verbal encouragements have now become a combination of human language and

24
Barad and Haraway both argue that scientific worldmaking is always the result of a process of 'becoming with' another. By referring to philosopher of science, Vinciane Despret in *Staying with the Trouble: Making Kin in the Chthulucene*, Haraway describes the 'becoming with' between birds and ornithologists as multispecies worldmaking (127–29).

animal calls, along with the practice of throwing supermarket sausage into the air towards the nets, thus overlapping several lifeworlds.

Once they are successful and a bird has flown into the net, a new phase in the sensory alignment with the Siberian jays begins. Now, a dexterous hand is necessary to free and register them. This requires general tactile skills and knowing specifically how to handle the birds, a kind of embodied knowledge that cannot be taught in universities; the actual handling of the birds can only be learnt through situated (sensory) enskillment in the field.

Barad describes the moment '[w]hen two hands touch [as one in which] there is a sensuality of the flesh, an exchange of warmth, a feeling of pressure, of presence, a proximity of otherness that brings the other nearly as close as oneself'.[25] Handling birds is thus an intimate, sensory experience unique with each bird. Indeed, the moment the biologists' hands wrap around the small bodies of the jays, there is a closeness that can rarely be encountered between *wild* birds and humans. In this moment, the natural world and the human world collide; generally, for humans, birds are unattainable creatures, out of reach; and, perhaps for this reason, they evoke significant fascination among ornithologists.[26]

However, while the birds have become the biologists' *prey* during the process of registration, the biologists also have power over the birds, independently of their intentions to turn them into prey or not. I suggest that the biologists are conscious of this and aware of their relationship to the birds. For this reason, these moments of catching and registering the birds create the most stress for them

25
Karen Barad, 'Diffracting Diffraction: Cutting Together-Apart', *Parallax* 20, no. 3 (2014): 153, https://doi.org/10.1080/13534645.2014.927623.

26
At this point, I would like to note that Michael and his team do not consider themselves ornithologists who only study birds. They consider themselves evolutionary biologists who work within Darwinian and evolution theory, attempting to understand how systems evolve and adapt to the environment (cf. field notebook entry, field season 2020, Day 6, during the drive back with Michael to pick up Marine and Camille).

during fieldwork. Their aim is not to harm the birds, nor to kill them or turn them into prey; instead, they want the birds to remain intact so that they become a valuable data source for their research.

While the biologists interact with the birds, they work hard to meet them at eye level. Holding them in their hands, they lift them up in front of their eyes, and the birds, with their deep black eyes, almost appear to be looking back. The biologists stroke their feathers and feed them sausage as a way of aligning with them instead of intimidating them. In this instance, they can feel how relaxed or stressed the birds are: some continue attempting to open their wings, whereas others try to defend themselves by pecking the biologists with their beaks, and some sit calmly in the biologists' hands with little resistance.

As much as the biologists attempt to treat the birds with respect by talking to them in a friendly manner while handling them and imposing the human world onto them, the boundaries of alignment between birds and biologists, human and non-human, become apparent. The catching of the birds is an ontologically violent act, as much as the biologists may defend it. It creates a duality in their sensory alignment: On the one hand, they attempt to empathise with the birds, while on the other, they must catch and register them to conduct their research and produce knowledge that will circulate between other humans. However, this knowledge may result in a higher awareness of the natural world and its systemic entanglements, and, thus, if mediated successfully, promotes its protection. The birds twitter and attempt to free themselves as they are being ringed, but the biologists do not respond. They hold onto them and continue their routine of measuring and collecting data from them.

It is clear in that moment that they are no longer response-able. This scenario is not a meeting between two subjects; instead, the birds have been turned into objects, thus establishing a subject–object divide anew. The birds cannot react according to their instincts, as they have been removed from their lifeworld and reterritorialised in the biologists' lifeworld, the research setting in which they are subjected to the biologists' will. Once the birds' weight, size, tail length, blood samples, and tail feather for DNA sequencing have been obtained, and the rings attached to their legs, they have literally become research *objects*. Henceforth, they will lead dual lives as research objects that provide the biologists with data, and subjects that continue their lives in the forest.

Sensory Alignment as Situated Practice

The biologists first undergo sensory enskillment in the field (Chapter 4). I referred to this as situated enskillment, based on Haraway's situated knowledges, to emphasise that this enskillment is highly dependent on the actual location, discourse, and thought style, and it is thus anything but a universal enskillment that applies to all situations. Sensory perception is always embodied and relates to a location: 'As such, perception is not an "inside the head" operation, performed upon the raw material of sensation, but takes place in circuits that cross-cut the boundaries between brain, body and world'.[27] As Willerslev states: 'Mimetic empathy, [...] does not imply therefore simply representation or imagination, but has a decisively corporal, physical, and tangible quality from which the former ultimately emerges and from which it derives its "material"'.[28] From this perspective, sensory alignment is a situated process that is embodied, cognitive, and in situ: in the field and in the biologists' bodies.

27
Ingold, *The Perception of the Environment: Essays on Livelihood, Dwelling and Skill*, 244.
28
Willerslev, 'Not Animal, Not Not-Animal', 648.

Sensory alignment with others, or other species, '[...] to a certain extent means that we "become" another kind "with" that being'[29] and 'becoming with is a practice of becoming worldly'.[30] I understand sensory alignment as a practice of 'becoming with', which leads to 'becoming worldly' and 'earthbound' (Latour). Therefore, I understand the notion of 'becoming worldly' as a version of situatedness that contrasts with Haraway's 'god's view', which would instead be something *heavenly,*[31] taking in a 'view from above'. Being *earthbound* suggests an opposition to the metaphysical notion of *heavenly* sciences, performed by a neutral scientist who is detached and does not become involved with the research object. With this, I consider the biologists 'earthlings' and thus entangled with their research subjects rather than detached from them. I argue that this entanglement becomes especially visible during sensory alignment.

Scientific work, as it is practised in my case study, and more generally, is always the result of a situated, partial perspective. However, this situated perspective cannot be considered independently of other relations that unfold in situ. Sensory alignment, as a combination of sensory skill, embodiment, and implicit knowledge, also contributes to this situatedness. Lastly, fieldwork occurs in the field, and as I have illustrated, is only possible through physical and sensory immersion. It is a practice that is literally situated – in this case, in the boreal forests of Arvidsjaur in northern Sweden – and is eventually part of situated mediations.

29
See Haraway, *When Species Meet,* 4, 16–17, as cited in Eduardo Kohn, *How Forests Think: Toward an Anthropology Beyond the Human* (Los Angeles: University of California Press, 2013), 140.

30
Haraway, *When Species Meet,* 3.

31
If one follows the metaphor of the (patriarchal) god view, which Haraway suggests describes objectivity, as it has been performed by the scientists thus far.

5.4. Situated Mediations in Ethnography

The concept of situated mediations can also be applied to my ethnographic work. During my first field days, as described in *Preparing* (Chapter 4), I learnt what tools and methods are best for conducting fieldwork with the biologists, what I can observe directly, where I need to ask for explanations, and what remains entirely invisible to me. I had to enskill my sensory competence on two levels: On the one hand, I learnt how to observe the birds, albeit less competently than the biologists, but still in such a way that I was able to follow their judgements and practices. On the other hand, I needed to learn how to observe the biologists during their fieldwork.

Where do I position myself the best? Do I attempt to align with their gaze when they look up at the sky through their binoculars, or do I simply study them while they do so? I extended my gaze in the same way they did with tools, such as my notebooks and camera. However, I had to develop a practice of mediating between my observations and documentations here as well. I needed to develop a system that helped me document and remember the field situation that would simultaneously serve as data. I had to learn which tool to employ in which moment and develop a routine for handling them in this specific field situation. At times I had to be fast, and at other times I had to face the cold and remove my notebook, take off my gloves, and start taking notes. Sometimes I had to carry my camera in one hand while skiing, hoping that I would not fall in the snow – which happened on several occasions.

The more I was immersed in the field, the less I employed tools, as the tool would always distract me from something else. Thus, I ended up mostly taking notes in my field notebook along with occasional photographs. However, both also worked as tools for framing and guiding my observations and selecting data points to store and take home. Even so, my data collection was less formalised than that of the biologists; in my case, they were part of the observational apparatus and could be considered situated mediations. Even more so than for the biologists, this situatedness was guided by my specific interest, disciplinary knowledge, and method of observation. My study is highly dependent on me, and I cannot simply be exchanged, as the (natural) scientific narrative of reproducibility claims. I am an inevitable and irreplaceable part of my apparatus of knowledge production. Replacing me means obtaining a different result.

Qualitative data collection also means deciding what situation is qualified to respond to my research interest and how to document this situation best to turn it into data. Unlike the biologists who predefine exactly what needs to be observed and what data must be collected, I follow a self-determined framework. However, I must still decide in almost every situation whether the information is relevant. And the *what* determines the way I may document it, thus turning it into data, which will later act as mediators between me and the field. While the biologists employ technologies for data collection as a means to an end without reflecting on the effects thereof and their role in shaping their data collection, I do the opposite. I continuously reflect on what way of documenting best serves the situation. Do I direct my camera at the biologists, which may cause them to behave differently, or do I simply take notes, bring them home, and immediately make an entry in my field diary, which I do every evening after fieldwork? In short: I reflect on the situations I create within my situated mediations during their employment. In this sense, I have developed routines for the employment of my data-collection practices, which allow me flexibility to react to the situations I encounter in the field. Simultaneously, I employ data collection as a framing practice to ensure that I do not collect arbitrary data.

Lastly, ethnographic research often follows grounded theory rather than a strict predefined research question. Thus, the process is iterative, whereas grounded theory is adjusted to the actual field situation, which can be observed. Along with this, the research questions are reformulated, and data collection depends more on the ethnographer's intuition. It is also based on sensory attention, experience, and a partial perspective, as well as mediations between field, discourse, and practice.

5.5. *Participant* Behaviour Observation

To consider multispecies observational apparatus from a methodological perspective, I conclude with a rather speculative play on the terminology of the formal methods of observation. In doing so, I aim to highlight the agencies of observation at stake. For this, I refer to my own participant observation, which I juxtapose with the behavioural observation study of the biologists.

Both research methods engage with observational settings and are thus bodily techniques that can be advanced and extended by technologies, as previously discussed. While one focuses on an observation method that engages with the research subject by participating, the research subject does not define the nature of the involvement, but instead describes the focus of the observation, the behaviour. In relation to what I have discussed earlier, I argue that, in both cases, a (sensory) involvement and (bodily) interaction between the observer and observed is occurring, even though these are based on different prerequisites. One is a human–human interaction between the biologists and me, the ethnographer, while the other is a human–non-human interaction between the biologists and the birds. Each interaction has different agencies.

I tentatively conclude by suggesting that the biologists' observations should be called *participant* behaviour observations. This term emphasises that the observation situation is not one that can be encountered in nature *as it is* but is one triggered by the biologists' manipulations of the birds' environment. Viewing the observational setting from this perspective also emphasises that – as discussed in the sections on sensory alignment – not only do the biologists have power over the birds, but the birds' behaviour also impacts the observations they trigger, and what can be turned into data and how.

This is where biology could learn from anthropology: The effect of the researcher's presence is something that is very well reflected in anthropology but still a blind spot in natural scientific research. Only recently, it seems, have scientists started to reflect on human involvement or experience as part of scientific knowledge production.[32] Ecofeminists such as Haraway and Barad have attempted to thematise this for decades by referring to knowledge production in the natural sciences (Haraway) and quantum physics (Barad), and how observational settings are a question of entanglements, as indicated by the well-known double-slit experiment in quantum physics.[33] I do not aim to question the objectivity of the data with this, but rather how it is constructed and rendered. My claim is in favour of a

32
Kohn, *How Forests Think*, 140.

33
The double-slit experiment, first performed by Thomas Young in 1801, illustrates the wave-particle duality of light and matter. When shooting light with one wavelength through a two-slit screen, the two resulting light sources create interference but still display wave behaviour on the other side of the screen. However, when replacing the light source with a proton, which, according to Isaac Newton, displays particle behaviour, the result becomes more complex. When shooting it through the screen, the protons also create an interference pattern, proving their wave behaviour. Accordingly, through the experimental setting, the physical state of the proton seems to have changed. In addition, when adding a detector on one side of the screen (which could be imagined as an observer) and turning this detector on, the protons display particle behaviour. However, the way they shoot through the two slits regularly alternates, which suggests an agency outside of human perception. Interestingly, when turning the detector off, the particles again display wave behaviour. Most surprisingly, when turning the detector on after the protons have been shot through the screen, they display particle behaviour as well, putting the (human) notion of linear time into question.

more self-reflexive and transparent scientific process that reveals the specifics of research subjects' transformation into data.

The notion of the biologists participating in the birds' activities for the brief period of data collection draws on the nature–culture divide, which is usually represented in the sciences. Illustrating how the lifeworlds of birds and biologists overlap during fieldwork allows for conceptualisation of a different kind of human–non-human relationship, one that departs from traditional notions of objectivity.[34] Rather than considering the scientist as neutral and external to the research setting, this shift in how the relationship is represented is necessary to shift the relationship between the natural world and the human world, as it is performed in the Anthropocene. Biologists, such as those in my case study, have the capacity, knowledge, and power to make these entanglements visible. However, they obscure those alignments in the results and thus also do not reflect on them. A shift is necessary in what is considered *objective* to create a shift in human–nature relationships, which are entangled rather than detached.

Lastly, I started this monograph by discussing the role of the visual as a practice of data collection from my design-informed perspective. And, indeed, it occupies a great deal of these practices. While my initial interest was sparked by the visual archival material I had access to, outlining data collection solely through the lens of vision – as the primary sensory practice – is ultimately insufficient. Accordingly, as I have attempted to discuss in this chapter, biological fieldwork is not only a matter of visual attention. It is the immersion in the field through holistic sensory perception, including the auditory realm and touch. The visual is a significant aspect that usually becomes evident in the results and which is also the most privileged sense in nat-

34
Vanessa Manceron, *Wild and Wonderful: An Ethnography of English Naturalists*, trans. Michael Taylor (HAU, 2025), https://press.uchicago.edu/ucp/books/book/distributed/W/bo239333429.html.

ural sciences and Western knowledge production. When examining the practices of data collection, scholars must extend their attention beyond the visual to other sensory and bodily practices to create more complex accounts of how knowledge is produced in the natural sciences. The following chapter describes the final stage of knowledge production practices as observed in my evolutionary biology case study.

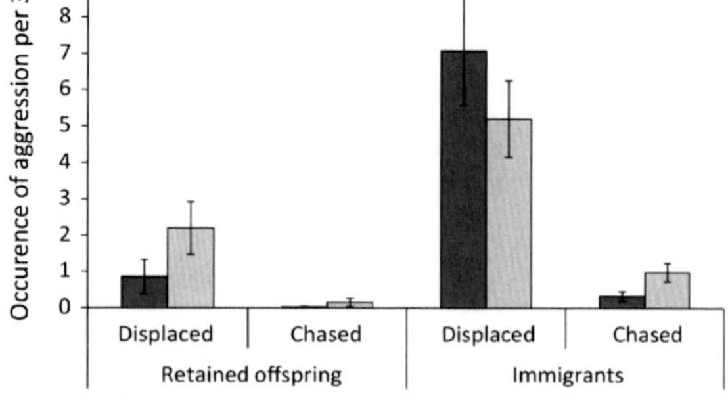

50 a

Fig. 3 Mean number of displacements and chasing events Siberian jay nonbreeders experienced during 30-min sampling observations (mean ± SE) in relation to nonbreeder sex and social relationship (retained offspring, immigrants). Female nonbreeders experienced more aggression than male nonbreeders, and immigrants experienced more aggression than retained offspring.

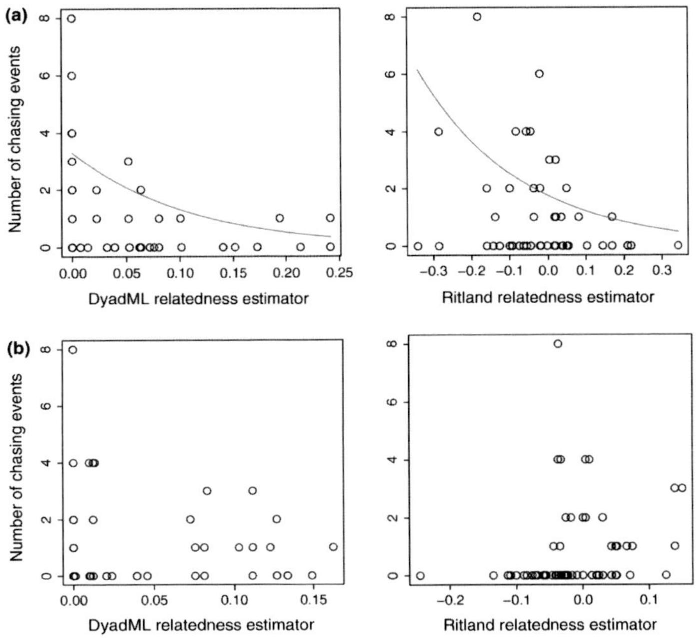

Fig. 4 Number of chasing events experienced by immigrants during 30-min sampling bouts in relation to (a) male breeder and (b) female breeder relatedness (DyadML, Ritland) between breeders and immigrants in the Siberian jay. More unrelated immigrants are exposed
50 b to significantly more chasing events.

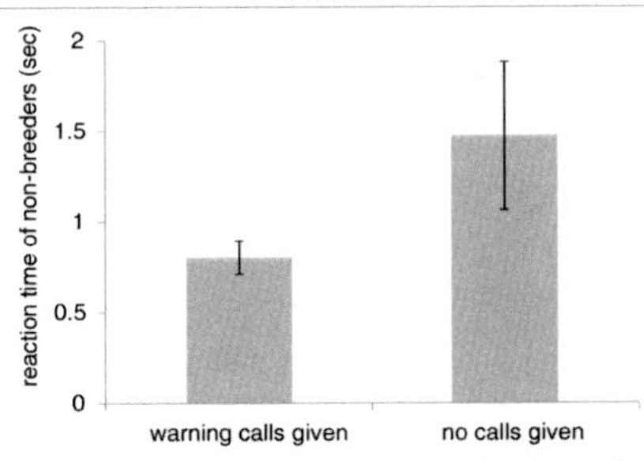

Figure 1 Effect of warning calls of Siberian jay breeders on the reaction time of non-breeders during simulated goshawk attacks. Non-breeders which received a warning call (N = 23 experiments) escaped faster to nearby cover than non-breeders which were not warned (N = 8 experiments). Data from two experiments were missing since the camera recording the whole set-up did not work properly.

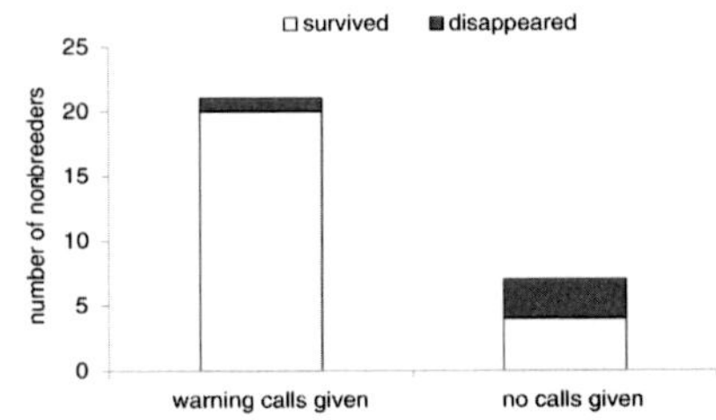

Figure 2 Influence of warning calls given by Siberian jay breeders during simulated goshawk attacks in autumn on subsequent winter survival of non-breeders (N = 28 individuals). Non-breeders living with breeders which warn (N = 21 individuals) had a higher over-winter survival than non-breeders which were not warned (N = 7 individuals).

51 a, b

Figure 50 a, b:
Screenshots of visualisations (graphs) of modelled data supporting the written account, from a paper published in *Molecular Ecology*.[35]

Figure 51 a, b:
Screenshots of visualisations (graphs) of modelled data supporting the written account, taken from a manuscript published in *Frontiers in Zoology*.[36]

35
Michael Griesser et al., 'Fine-Scale Kin Recognition in the Absence of Social Familiarity in the Siberian Jay, a Monogamous Bird Species', *Molecular Ecology* 24, no. 22 (2015): 5726–38, https://doi.org/10.1111/mec.13420.

36
Michael Griesser, 'Do Warning Calls Boost Survival of Signal Recipients? Evidence from a Field Experiment in a Group-Living Bird Species', *Frontiers in Zoology* 10, no. 1 (2013), https://doi.org/10.1186/1742-9994-10-49.

Chapter 6:

PRODUCING

The environment in which data were collected during the two field seasons of the spring of 2015 and 2020 is complex and multidimensional. During fieldwork, the biologists and I were immersed in the habitat of the Siberian jays in the boreal forest of Arvidsjaur in Sweden. Here, the birds live among trees, bushes, rivers, bridges, marshlands, paths, mushrooms, lichen, moss, and snow. They witness the different seasons and are exposed to the environment. Occasionally, they fall prey to the surroundings with which they are entangled. The trees serve as spaces to build nests and raise offspring under the protection of the branches, while the berries, seeds, nuts, worms, spiders, and insects serve as nutrition. In winter, the jays become hoarders and store their food in the bark of the trees; one bird remembering several hundred hideaways. In summer, the forest protects them from predators, but in winter, the brown-orange birds stand out against the white snow, making survival more difficult, especially in scarcely vegetated or deforested landscapes.[1] Aspects of this multiplicity and ecological entanglement become visible as material traces in the raw data. In the field notebooks, one can see the marks of raindrops that had fallen onto the pages and the remnants of midges that could not escape before the notebook was shut.

1
Layton-Matthews, Ozgul, and Griesser, 'The Interacting Effects of Forestry and Climate Change on the Demography of a Group-Living Bird Population'.

Therefore, without further transformation, formalisation, and abstraction, certain characteristics of the birds cannot be converted into formal scientific results that contribute to knowledge production – an issue closely tied to the loss of their environment. While on-site in Sweden, the birds' complex entanglements with the forests can be witnessed and explored; for scientific purposes, the birds must be disentangled and extracted from the forest by the biologists and transferred to a different environment. They must be transferred to offices where the actual knowledge production occurs. Only here, as a result of data processing, do *insights* appear to be gained and *facts* produced on the behaviour of the Siberian jays.

The biologists return from the field to their offices with notebooks filled with handwritten charts, notes on separate sheets of paper, and printed maps to organise fieldwork. In addition to their notebooks, they have SD cards containing 15-minute video recordings of the birds' behaviour, and blood samples, tail feathers, and body measurements of the new birds that have become research objects. These are raw data, resulting directly from the field, mostly produced based on formalised empirical observations of the Siberian jays. To produce knowledge, these data must be processed.

This processing creates the gap I mentioned in Chapter 1 of the book. The gap is defined by the invisible steps between bird and published paper. It is marked by a (visual) difference between the representation of data as published results and the raw field data that are not made public. The raw field data are produced in a technically simple manner during fieldwork, often by means of notes on a video recording. The visible results are highly technical images generated at the end of the research process, based on data modelling, coding, and algorithms applied to raw data plots (Figures 50 and 51).

These final technical images are 'not simply compendia of visible information, they document the schematic presentation of carefully distilled and edited observation',[2] hiding the technical and other operations that led to them. The visualisations usually consist of a few lines on a graph with an *x*- and a *y*-axis, along with abstract numbers, individual letters, and a few indexical words that provide some details of the situation, such as 'displaced', 'chased', 'male nonbreeder', or 'female nonbreeder'. Colour is used solely to distinguish the data. The life cycles of most data that led to the refined images have become invisible in these datasets. The data have become visually generalised and homogenised and have lost their complexity. They serve as visual proof of the arguments made in the paper.[3]

This chapter focuses on the life cycle of data based on the archival material of my case study, along with the knowledge and experience of data processing gained through fieldwork and conversations. I consider mainly how the datasets are prepared for publication rather than the process of writing a scientific paper. I describe the operations that the biologists apply to the raw visual material from the field to produce a *robust* scientific result, following the threads that lead from bird to paper and from inscription to inscription.[4] By such means, I return to Latour's concept of creating reference in the scientific process. While Latour argues that the individual steps from research objects to results are shaped by a 'long series of manipulations' to 'narrow the gaps',[5] which will later be made invisible, I suggest changing the metaphor.

2
Johanna Drucker, 'Graphesis: Visual Knowledge Production and Representation', *Poetess Archive Journal* 2, no. 1 (2011): 7, https://paj-ojs-tamu.tdl.org/paj/article/view/4.

3
Johanna Drucker, *Graphesis. Visual Forms of Knowledge Production* (Cambridge, MA: Harvard University Press, 2014); Latour, 'The More Manipulations the Better'; Merz, 'Bildkomplexe als Geschichten: Naturwissenschaftler Erzählen'.

4
Latour, 'The More Manipulations the Better'; Latour and Woolgar, *Laboratory Life*.

5
Latour, 'The More Manipulations the Better', 348.

Accordingly, I introduce the concept of filters rather than *manipulation* to describe these practices and provide a different understanding. The concept of filtering emphasises two aspects: First, it poses the question of what acts as a filter and how. Second, it highlights who applies these filters that facilitate the scientific practice (situated mediations) and help to make agencies visible. When examining raw data rather than the final object, this situatedness becomes visible in material traces that the biologists do not consider data. Attending to these emphasises 'a performative understanding of the scientific practice'.[6] Understanding the process of transforming raw data into final results as one of filtering helps illuminate how biologists conceptualise and develop their research object.

While the term 'manipulation' has a negative, slightly moralistic, connotation, and could perhaps imply deceit, the term etymologically means the handling of persons and/or objects by hand, often to one's own advantage. The Latin word *manus* means hand; thus, all other body parts are excluded. However, as I have discussed in previous chapters, the biologists' entire bodies and all their senses are involved in fieldwork. The term *filtering* does not have this negative connotation. I consider it wise to use a metaphor that does not inherently refer to any particular body part and does not privilege specific body parts over others. This allows me to discuss the practices without simultaneously evaluating them.

In the following section, I discuss *filtering* as a metaphor for the practices that help transform the birds into scientific output. Here, I refer to theories on different media modalities, how they constitute themselves in their usage, and how they are made visible and/or invisible. In particular, Offen-

6 Barad, *Meeting the Universe Halfway*, 49.

huber's[7] concept of *autographic data* helps shift attention to the *side*-data that become visible in the raw biological data. Furthermore, Jane Bennett's theory on the *vitality* of things offers an understanding of these data as having agency rather than being biological waste products.

Having outlined the steps of filtering (6.1), I describe how the working environment shifts from field to office, and how this is closely related to *Relocating Birds and Biologists* (6.2). I conclude the chapter with a *Thick Description by means of Visualisation* (6.3), referring to the ethnographic method of thick description and the concept of *infrastices,* in which Tim Ingold and Mike Anusas state that interfaces should be transparent.

7
Dietmar Offenhuber, 'Data by Proxy – Material Traces as Autographic Visualizations', *IEEE Transactions on Visualization and Computer Graphics* 26, no. 1 (2020): 98–108, https://doi.org/10.1109/TVCG.2019.2934788; Dietmar Offenhuber, 'Dis/Entangling Perspectives in Material Research', in *Data Autographies – A Material Perspective on Data Visualization and Evidence Construction* (online: Matter of Activity: Humboldt University Berlin, 2022), https://www.matters-of-activity.de/en/activities/6386/dis-entangling-perspectives-in-material-research.

52

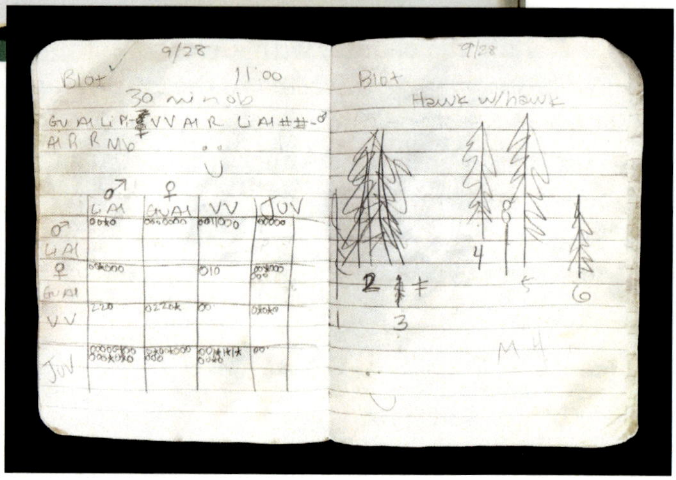

53

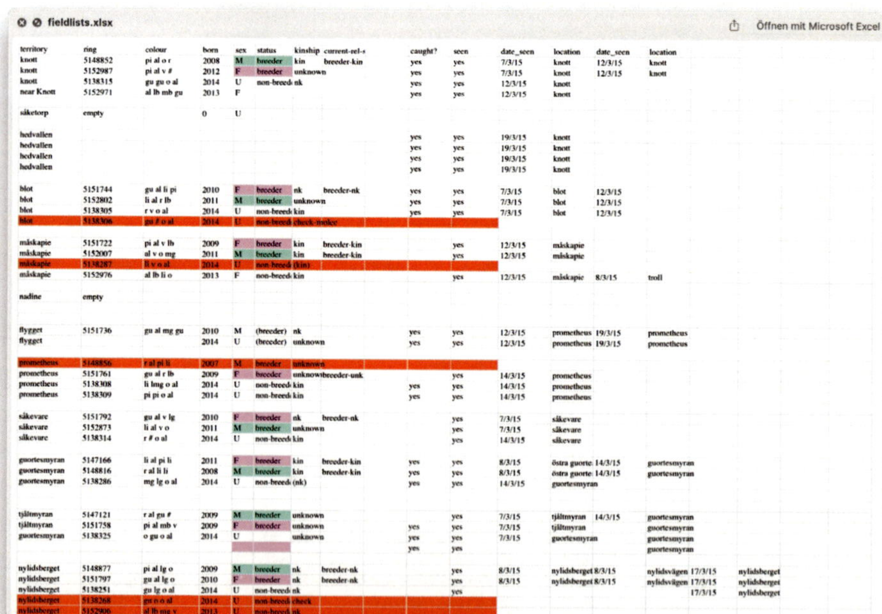

territory	ring	colour	born	sex	status	kinship	current-rel-s	caught?	seen	date_seen	location	date_seen	location		
knott	5148852	pi al o r	2008	M	breeder	kin	breeder-kin	yes	yes	7/3/15	knott	12/3/15	knott		
knott	5152987	pi al v #	2012	F	breeder	unknown		yes	yes	7/3/15	knott	12/3/15	knott		
knott	5138315	gu gu o al	2014	U	non-breed	nk		yes	yes	12/3/15	knott				
near Knott	5152971	al lb mb gu	2013	F				yes	yes	12/3/15	knott				
såketorp	empty		0	U											
hedvallen								yes	yes	19/3/15	knott				
hedvallen								yes	yes	19/3/15	knott				
hedvallen								yes	yes	19/3/15	knott				
hedvallen								yes	yes	19/3/15	knott				
blot	5151744	gu al li pi	2010	F	breeder	nk	breeder-nk	yes	yes	7/3/15	blot	12/3/15			
blot	5152802	li al r lb	2011	M	breeder	unknown		yes	yes	7/3/15	blot	12/3/15			
blot	5138305	r v o al	2014	U	non-breed	kin		yes	yes	7/3/15	blot	12/3/15			
blot	5138306	gu # o al	2014	U	non-breed	check-molee									
måskapie	5151722	pi al v lb	2009	F	breeder	kin	breeder-kin		yes	12/3/15	måskapie				
måskapie	5152007	al v o mg	2011	M	breeder	kin	breeder-kin		yes	12/3/15	måskapie				
måskapie	5138287	li v o al	2014	U	non-breed	(kin)									
måskapie	5152976	al lb li o	2013	F	non-breed	kin			yes	12/3/15	måskapie	8/3/15	troll		
nadine	empty														
flygget	5151736	gu al mg gu	2010	M	(breeder)	nk		yes	yes	12/3/15	prometheus	19/3/15	prometheus		
flygget			2014	U	(breeder)	unknown		yes	yes	12/3/15	prometheus	19/3/15	prometheus		
prometheus	5148856	r al pi li	2007	M	breeder	unknown									
prometheus	5151761	gu al r lb	2009	F	breeder	unknown	breeder-unk		yes	14/3/15	prometheus				
prometheus	5138308	li lmg o al	2014	U	non-breed	kin		yes	yes	14/3/15	prometheus				
prometheus	5138309	pi pi o al	2014	U	non-breed	kin		yes	yes	14/3/15	prometheus				
såkevare	5151792	gu al v lg	2010	F	breeder	nk	breeder-nk		yes	7/3/15	såkevare				
såkevare	5152873	li al v o	2011	M	breeder	unknown			yes	7/3/15	såkevare				
såkevare	5138314	r # o al	2014	U	non-breed	kin			yes	14/3/15	såkevare				
guortesmyran	5147166	li al pi li	2011	F	breeder	kin	breeder-kin	yes	yes	8/3/15	östra guorte	14/3/15	guortesmyran		
guortesmyran	5148816	r al li li	2008	M	breeder	kin	breeder-kin	yes	yes	8/3/15	östra guorte	14/3/15	guortesmyran		
guortesmyran	5138286	mg lg o al	2014	U	non-breed	(nk)		yes	yes	14/3/15	guortesmyran				
tjåltmyran	5147121	r al gu #	2009	M	breeder	unknown			yes	7/3/15	tjåltmyran	14/3/15	guortesmyran		
tjåltmyran	5151758	pi al mb v	2009	F	breeder	unknown		yes	yes	7/3/15	tjåltmyran		guortesmyran		
guortesmyran	5138325	o gu o al	2014	U		unknown		yes	yes	7/3/15	guortesmyran		guortesmyran		
								yes	yes				guortesmyran		
nylidsberget	5148877	pi al lg o	2009	M	breeder	nk	breeder-nk		yes	8/3/15	nylidsberget	8/3/15	nylidsvägen	17/3/15	nylidsberget
nylidsberget	5151797	gu al lg o	2010	F	breeder	nk	breeder-nk		yes	8/3/15	nylidsberget	8/3/15	nylidsvägen	17/3/15	nylidsberget
nylidsberget	5138251	gu lg o al	2014	U	non-breed	nk			yes					17/3/15	nylidsberget
nylidsberget	5138268	gu o o al	2014	U	non-breed	check									
nylidsberget	5152906	al lb mg v	2013	U	non-breed	nk									

54

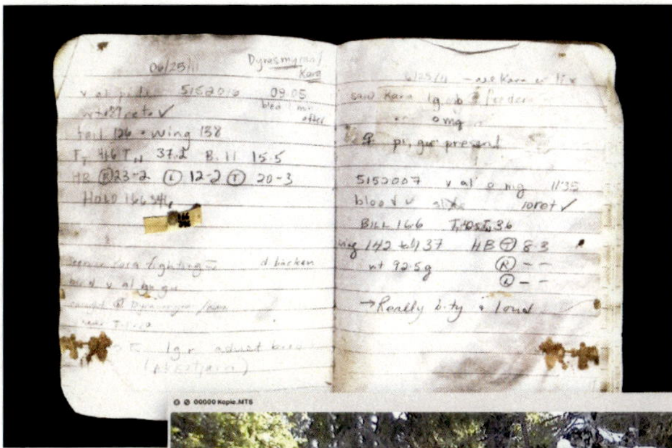

55

56

57

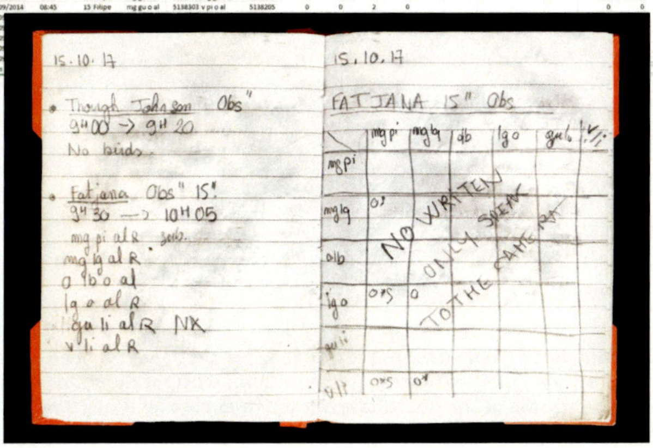
15.10.17
Though Johnson Obs"
9H00 → 9H20
No birds.
Fatjana Obs" 15"
9H30 → 10H05
15.10.17
FATJANA 15" Obs
NO WRITTEN
ONLY SPEAK
TO THE CAMERA

58

Figure 52:
Field notebooks stored in a cabinet in Michael's office at the University of Zurich. Zurich, 2018.

Figure 53:
Screenshot that I took after scrolling through a PDF file with scans of the biologists' field notebooks. The screenshot shows a double page of Sinja's field notebook with a behavioural observation protocol in territory *Blot* to the left and side-notes to complement the video recording on the right.

Figure 54:
Excerpt from field list for the study in 2015 (Excel chart) representing the territories and the Siberian jays living in them. The breeding couple and Siberian jays that are missing are highlighted. Zurich, 2021.

Figure 55:
Screenshot of a double page of a field notebook from 2011 with bodily measurements of Siberian jays taken by biologist Lena in *Kara* and *Dyrasmyran.* A radio tag is taped onto the page on the left. Several brown marks indicate blood or dirt. Zurich, 2018.

Figure 56:
Screenshot from video recording by Kate in territory *Knott.* Arvidsjaur, 2015.

Figure 57:
Screenshot of Excel data plot with behavioural observation after transfer from video in the office. Zurich, 2018.

Figure 58:
Double page of Jasmin's field notebook from 2017. On the left showing that she had been to the territory *Tough Johnson* where she waited for 20 minutes without Siberian jays appearing. Below is an indication of the time she spent in *Fat Jana* (9:30–10:05 a.m.) and the six Siberian jays she identified. Behavioural protocol of the Siberian jays in territory *Fat Jana,* which Jasmin started but then omitted: 'No written, only speak to camera.' Zurich, 2018.

6.1. Filters

Filtering is a mechanical process that separates or *purifies* substances. During the process, the *input* is separated into at least two components, namely *residue* and *filtrate*, determined by the permeability of the filter. For my case study, I suggest conceptualising the field as the input and the raw data as the filtrate. The residue is what remains in the field: all formal and informal observations that did not become data. In the next step, the data that have been transported to the researchers' offices become the input, while a *refined* version thereof is the filtrate. Consequently, the aspects that are unimportant to the biologists, such as side-notes in their field notebooks, become the *residue*. In each of the steps, the permeability of the filter is the interest with which the biologists study the input (data). Based on this, they decide what should be kept and what should be discarded. The permeability determines which information passes through the filter and becomes the *filtrate*, and which information becomes *stuck* in the filter as the *residue*. The tools, such as the behavioural observation protocol, video camera, and computer interface, support and determine the individual steps of filtering. Thus, together, the biologists and their tools form a filtering apparatus.

The filters have different levels of permeability. In some cases, the filtrate does not carry any remaining sediments that might provide insight into the conditions of its production, such as the refined Excel sheets on computer screens (these differ from those on which the biologists manually captured information during the fieldwork season). In other cases, such as the video recordings or behavioural observation protocols, additional information becomes visible as autographic data. The filters, with their varying permeability, can thus be separated into densities, ranging from thick to thin, or obscure to transparent.

Again, during these practices, the biologists do not act as neutral bodies. The filtering is based on their intra-action with the technical operations, discourse, and apparatus of knowledge production that define their decisions. In short, the filtering practices are another form of situated mediations, which are a combination of situated enskillment, formalised training, discourse literacy, skilled visions, and the employment of technologies, with which the biologists filter aspects of the birds. Thus, through the steps of filtering, three kinds of situated mediations become visible.

The process of filtering occurs in three main steps: from bird to raw data (in the field), from raw data to analytical data (in the office), and from analytical data to a graphical representation as the result of the scientific process. The raw data collection is based on a first filtering process of skilled mediations as filters. The second is from raw field data to analytical data by capturing the raw data on an Excel sheet where they are combined with all the data from the previous seasons and homogenised. The third filter is applied to the Excel sheet, based on algorithms, where specific aspects of the dataset selected by the biologists are quantitatively combined and modelled into a final visual representation of the results for publication along with a written account. In most cases, this is a graph or schematic diagram, as described at the beginning of this chapter. Each of these steps is shaped by the tension between the visible and invisible. To make one aspect visible, something else must often be removed from the biologist's sight, as it prevents them from *seeing* the facts.

It is important that nothing is discarded, even if all data points are not transferred. Waste data can be stored in the archives, along with the notebooks, individual sheets of paper, and hard drives containing the video recordings. From a biological perspective, this is a process of refining data by separating it. However, to guarantee objectivity and ensure that the data are not arbitrary, certain organising principles must remain constant. Accordingly, in these filtrations, the bird colour codes (IDs), territory names, and study sites, among other things, circulate as 'immutable mobiles', which are '[...] objects which have the properties of being mobile but also immutable, presentable, readable and combinable with one another'.[8] From an STS perspective, the data transform from a thick description that is rich, multi-layered, and complex to a *flattened* result where the data have lost their complexity. Therefore, I suggest focusing on the *residue* to reintroduce the practices that produced the scientific results into the discourse. As outlined in the introduction, I filter them back *in*to the knowledge of the scientific process. I do this by mixing the *poor*ified[9] filtrate with the residue, thereby complexifying the data again.

To conceptualise the residue, I refer to 'autographic data'. Offenhuber introduced this term in *Data by Proxy,*[10] and it will help describe the 'surplus data' of scientific knowledge production. Offenhuber understands autographic visualisation as a graphical practice of data visualisation that specifically attends to the material entities of data and the self-inscription of data on these entities. He refers, for instance, to tree rings as data visualisation, where material and data have become inseparable.[11]

8
Latour and Woolgar, *Laboratory Life,* 7.

9
*Poor*ified as in 'purified': I allow myself this play on words to emphasise the duality of the data. From a design-informed STS perspective such as mine, the data that remain after processing become poor in the sense that they lose complexity. They become 'flat' – as I call it – for the sake of reductionism. However, from the biological perspective, they become *pure*ified, which is an essential step for the biologist to *see* their proof by presenting their data visually with graphs.

10
Offenhuber, 'Data by Proxy – Material Traces as Autographic Visualizations'.

11
Cf. ibid., 99.

Autographic visualisations are defined as a process in which 'phenomena [...] reveal themselves as visible traces'.[12] These 'epistemologies of traces'[13] decentre the human, as they can inscribe themselves on the humans and other carriers of data traces. As these traces are usually filtered out during subsequent steps, they are disregarded by scientists. In this sense, autographic data materialise and inscribe themselves on the data carriers, for instance in the biologists' notebooks. They provide insight beyond the data collected, for instance, the processes, practices, and environmental aspects that accompany the research process. Other than the tree-ring example, they can be separated by filtering, as I shall illustrate.

6.1.1. First Filter: Transforming Observations into Raw Data

When studying the behaviour of the Siberian jays, bare visual observation is insufficient, as discussed in the previous chapter. Therefore, the biologists' observations in the field are captured by behavioural observation protocols and video recordings. These practices function as filters that sift the relevant information from all the information the biologists receive. The behavioural observation protocol works as a filter by predetermining what should be documented and turned into data (Chapter 5).

Observations are considered valid data only when they are visualised and stored on data carriers as raw data – the filtrate produced by the behavioural observation protocol. Without documentation, there would be no scientifically valid data to be systematically analysed and interpreted. Regardless of whether a study is in ethnography or biology, data are inevitable to produce scientific output. Therefore,

12 Ibid.

13 Cf. Offenhuber, 'Dis/Entangling Perspectives in Material Research'.

the life cycle of data starts with this initial *filtrate,* the documentations made in the field based on situated mediations, which, as illustrated earlier (cf. Chapter 5: 5.2), work as filters during fieldwork. They isolate the birds from the field, thus reducing their complexity. The birds are slowly turned into abstract scientific representations through this process.

Along with organising principles (such as naming the study areas, bird families, and individual birds, which circulate as 'immutable mobiles' (Latour) through the datasets and allow consistency and connection between the specific birds and the collected data), and formalised methods (that enable a steady collection of data), the biologists formalise quantitative theories and produce scientific evidence. The territory names function as references to the groups and bird territories, and in terms of data analysis, they allow the biologists to disregard the physical and geographical relations of these territories with the others in the study area, as they are replaced by cells on an Excel sheet. The field data contain much more than these organising principles and the information collected through formalised methods. Besides the relevant biological data, the notebooks and raw field data also carry material traces and autographic data.

In this sense, information has both been extracted from the birds during data collection and been added to. The 'adding', from the biologists' perspective, occurred as a side effect, as in the case of their handwriting. For the filtering process, this additional autographic data is *noise* that must be separated from the data. For now, I focus on this noise before it is filtered out in the next steps of the research process.

To follow the material traces of what becomes visible in raw data, I make use of my archival material. I open the folder named 'Material' on my MacBook, followed by the subfolder 'GSA_FS18_Labday_04052018', where I find the PDF called 'Fieldnotes_Notebooks_students_2011-17'. It contains a collection of several field notebooks from master's, PhD, and postdoctoral students who conducted fieldwork in Arvidsjaur between 2011 and 2017. I scanned these notebooks one day in the researchers' office.

Michael had granted me access to his filing cabinet (Figure 52), which usually remains locked. He prefers that the material remains in the office, but I can go through everything and make notes and scans, which I eventually did with my smartphone because there was only one (very slow) scanner in the office. Therefore, I did not scan each notebook entirely, but instead focused only on excerpts that struck me. I focus on differences in the noting practices: aspects that did not appear to be part of the process of biological data collection, such as small drawings or traces of the research conditions inscribed on the pages of the notebooks, such as dead midges or pages wrinkled from the rain.

Scrolling through these excerpts now, I stop on page 64 of 73: an excerpt of biologist Sinja's notebook from 2014 (Figure 53). I have never met Sinja, but I can identify with her by examining this page of her notebook. I imagine how she may have worked with the birds. To follow her thoughts, based on this archival material, I study not only the pages in the notebook but also the material relating to her notes, such as the Excel sheet of all the registered birds. I attempt to understand how many birds she managed to attract, determine in which territory and study area she worked, and learn the characteristics of her environment. Based on her notebook entry, I trace Sinja's observational process – her situated mediations – to understand how she employed the behavioural observation protocol in the field.

The scan reveals that the notebook itself is somewhat weathered on the edges, likely from its frequent use during fieldwork, and as far as I can see on my scan, the right side of the notebook became wet at some point, as the edges have a weathered appearance and some of the lines have rubbed off. Before I study the notes more specifically, I am struck by Sinja's handwriting, which is quite large compared with that of others. There is more space between her capital letters, which makes it easier to identify her notes and helpful for me as an outsider wanting to study and understand the relationship between the biologists, how the birds are studied, and the role of the notes. I can also study the density of her lines by the varying shades of grey of the graphite pencil on the paper.

At the top of the left-hand page of the behavioural observation protocol, she wrote 'September 28th', the name of the bird territory, '*Blot*', the time, '11 o'clock', and the task she fulfilled, a 30-minute observation,[14] which she noted as '30 min obs'. There are other elements to her notes as well. I can see moments of hesitation where she crossed out the male symbol and replaced it with the female symbol at the top of the left-hand page. I feel as though I can also see how she hesitated for a moment when drawing the trees on the right-hand side page. The fact that she drew a '2' over a '1' leads me to believe that she changed her mind as she was numbering the trees (to guide the biologists who will eventually analyse the video recording).

Looking at the handwritten protocols allows me to examine the boundaries of the visual filtering practices. First, for the biologists, the behavioural observation protocol is a highly efficient and effective tool to monitor and document the complex aspects of social behaviour among several birds. It combines time, three-dimensional movement, and different kinds of interactions into one simple two-dimensional visualisation that usually fits on half a page of an A6 notebook. With this technique, the biologists can translate a 15-minute (or longer) excerpt of an event they triggered in the field among several individual birds into an immutable mobile. The result is a document indicating the location, bird territory, date, time, and IDs of the birds present during the behavioural observation protocol, as well as the interactions between individual birds, which are quantified.

14
During my study, the video recordings were extended from 15 to 30 minutes to ensure that all relevant behaviour could be observed.

The birds' IDs are in the vertical and horizontal columns with boxes containing the relevant codes for the individual interactions observed. This immediately allows an (en-) skilled eye to gather information on the individual birds. It functions as a condensing tool and a filter that captures only the relevant interactions and events that occurred during this period. It also helps to turn qualitative observations into quantitative outcomes that provide the biologists with clear information on their research object.

The field data do not provide any direct information on anything else, such as the environment, weather, temperature, other animals, or predators. Nor do they indicate any human conditions and interventions; the disposition of the biologists; infrastructure, such as houses, cars, railways, paths, forestry works; or other people, such as the Sámi reindeer herders, who occasionally drive through the forests on their snowmobiles. However, this information is not entirely filtered out. In some cases, the biologists note that heavy forestry work has occurred, that they spotted predators, or events occurred that may influence the birds' behaviour. According to Sinja's observation, this was not the case.

On further examination of the double page, I notice four other marks that I can only interpret as two 'u' shapes with two dots on top; in German they could be read as the umlaut (ü). To me, it seems more likely that these shapes represent smiley faces. Considering the location at the bottom of the double page, I do not know whether I should interpret this as Sinja's satisfaction with a successful observation, or a secret code she is communicating to the biologists who will process the data. It could also function as a mnemonic technique to help her recall the field situation when she uses the notebook again later, for instance, when choosing a suitable group of birds with which to conduct further observations.

There is also a combination of the letter M and the number 4. I do not know what it represents and I disregard it for now. Lastly, I notice a small hook in green ink on the top-left corner of the page. I assume that the biologists who transferred the notes (along with the relevant video recording) to a digital datasheet had written this. To recall specifically where *Blot*, the territory in which Sinja worked, is located, I refer to a document called 'Maps_2020.pdf'.[15] It contains maps of the three study areas. When Sinja made her observations in 2017, the study site was different. Some of the birds have since died, perhaps from the cold winter, predators, or age, and some have emigrated. Some territories did not yet exist, while others have since disappeared. As I have access only to the maps of 2020, I can see some of these changes on the maps, while I must assume others based on the material. I find *Blot* on the map for *Managed* in the top-right-hand corner and disregard the maps for *Fat Road* and *Reivo* for now.

I subsequently move on to a different file called 'fieldlists.xlsx' (Figure 54), an Excel datasheet from the final data plot, which I received from Michael for my fieldwork in 2015, where I also look for *Blot*.[16] On the left page, I read 'GU AL PI', followed by the male symbol, which has been crossed out and replaced by the female symbol, followed by a combination of letters and signs ('VV AI R Li AI ##'), and then the male symbol. The letters represent abbreviations for the Swedish words for colours, an additional translation step the researchers must perform. This has now changed and since 2023 they employ the English translation to make it easier for the international community of researchers involved in the project. I find another combination in the next line ('AI RR Mb'). Comparing this with the Excel field list from 2015, I can confirm that there are four birds in this territory; the code 'VV AI R Li AI ##' indicates the individual colour IDs of two birds, which can be separated into 'VV AI R' and 'Li AI ##'. The latter bird has lost two

15
Michael asked me not to publish detailed representations of the territory maps to avoid revealing the exact locations of the birds. Figure 19 provides a general idea of what the maps look like.

16
Both a copy of the map and the field list were likely glued into Sinja's notebook in the front and back, as was done by all the biologists to have these essential details close to hand at all times. I did not scan the list and the map, as they are in every notebook I studied.

rings since 2015 as indicated by the '#'. I can derive from the Excel sheet that the missing IDs were red and light blue. The other two birds are no longer the same in the Excel sheet and notebook. Presumably, they have changed territories between 2015 and 2017, which is likely given they were both non-breeders, who usually find a new territory in which to start their own group. The birds that remained the same are the breeding couple (male and female).

As can be seen with Sinja's corrections, media have their obstinacies, which become particularly visible in moments of interference. The reason Sinja needed to make corrections is unknown to me, but, as media scholar Judith Willkomm points out, this moment of interference makes the modality of the notebook visible.[17] While most media are usually hidden and naturalised, through this correction, the notebook itself becomes visible.[18] This 'constitutive moment'[19] illustrates how a notebook is different to, for instance, digital applications as actants (to echo Latour) in a research setting. From my perspective, the use of a handwritten notebook allows me to trace the moment of data collection in richer detail than would be the case for a digital recording.

With the notebook, corrections cannot simply be made if Sinja does not have an eraser to hand, which the biologists usually do not. Thus, her error becomes a visual trace that tells its own story. This kind of *storytelling* is a result of the modality of the notebook. In another case, if the notebook had been replaced by an application for conducting the behavioural observation protocol – which exists but has not been used by Michael and his team thus far – the error would have been invisible and remained a brief, untold moment that Sinja experienced in the field.

17
Translated by Judith Willkomm, 'Mediatisierte Sinne und die Eigensinnigkeit der Medien. Für eine Medientheoretische Sensibilisierung der Sinnlichen Ethnographie', in *Ethnographien der Sinne*, ed. Lydia Maria Arantes and Elisa Rieger (Bielefeld: Transcript Verlag, 2014), 51.

18
Ibid.

19
Latour and Woolgar, *Laboratory Life*, 88.

While the chart on the left-hand side of the field notebook (once it has been completed) delivers initial insights into the behaviour of this group of birds, the drawings on the right represent the setting and complement the video that Sinja has been recording during the 30-minute observation, as is usually the case. The numbers of the trees help systematise the verbal descriptions, making them more precise by giving the trees designations to which Sinja can refer. In addition, the small vertical line with the two horizontal lines across indicate where the feeder has been located, while the line at the top right with the two circles on top between trees 4 and 5 indicates where an artificial hawk was positioned as part of the experimental setting.

The resulting behavioural observation protocols no longer show the actual birds, but indexically refer to them based on their colour IDs, which consist of codes that relate the (existing) data points to the individual birds. The birds have thus been completely replaced by their IDs. This code remains the same across the field notebook, maps, and Excel sheet. Thus, it provides an overview, enables the biologists to monitor the bird population (which is only possible through this reference system they have developed), and helps them organise their research and maintain consistency. In the protocols, the birds themselves are filtered out, along with the landscape and the experimental setting with the feeder. They have been removed from their context during their transformation from bird to paper. What remains is a cryptic chart that, to the expert, provides insights into the birds' interaction around the feeder. No other information remains in these documents. Certain interactions between individual birds become visible, which would hardly be possible without this documentation. They would have remained an inseparable meshwork of interactions between birds in the forest that would not have been identifiable to the biologists as the birds of a specific territory or group, as it is incredibly difficult to tell them apart without their IDs.

These data will later partially be turned into digital datasheets and in this way prepared for further processing and analysis. It partly supports the video processing that comes next in the biologists' analysis. However, Sinja's side-notes and the drawing will not be transferred; they only help organise the fieldwork and its analysis for the peer biologists who will analyse the raw data. Her 'errors' will not find their way into the next dataset. Subsequently, the biologists who transfer the data check the side-notes to see if their colleagues in the field documented any irregularities that may have influenced the data collection, which they may need to consider when processing the data.

Three Categories of Data

I suggest that there are three categories of filtered data based on the information appearing on the double page. First, the biologically relevant data are collected based on situated mediations, as they become visible in the behavioural observation protocol in the video recordings or side-notes in the field notebooks. The raw biological data contain two levels of autographic data as material traces, thus forming a second category that inscribes itself on the data carriers as part of the data-collection process. These traces include, for instance, how the behavioural observation protocol is developed, the biologists' handwriting, and their errors (as in the case of Sinja), or the personal methods employed as mnemonic techniques to support the observational tasks. For instance, Marine and Camille first translated the colour codes from Swedish to English or French before they made their observations, while other biologists copied the colours directly to identify them more easily. As a third category, there are material traces that are not part of the actual documentation itself but result from the conditions and the environment of data collection, such as the weathered notebooks, marks from raindrops, or dead midges. In other words, they offer insight into the conditions in the field.

In biologist Lena's field notebook from the summer of 2011 (Figure 55), I can see traces of all three categories. The pages have many folds and the edges are weathered. The notebook looks as though it became wet, which suggests it was raining during fieldwork. There are also brown traces of dirt or blood on the pages, which likely stem from Lena measuring the birds.

While these are autographic data, there is a (category 1) side-note indicating 'Really bity + loud' on the right-hand page, probably characterising the juvenile bird that Lena ringed and measured in this territory. In examining these pages, I wonder whether some of the blood that Lena sampled from the bird was accidentally transferred to the notebook, or if it is her own blood from a bite by the aggressive bird. However, I can only speculate; perhaps the brown marks are simply dirt.

Out of curiosity, I quickly check the field list from my previous trip in 2020, which is glued into my notebook, to see whether the bird is still alive. And, indeed, I find the bird; however, it is mentioned under a different territory – it now lives in *Måskapie*. According to the field list, it lives there as the male breeder along with a female breeder and a third bird, which may have changed territories again by now.

Studying the PDF with the territory maps also allows me to geographically locate the study, and I can start looking up certain locations to see where exactly the biologists worked and what their surroundings were. These geographical references, territory names, and bird codes usually disappear in the final papers, as the maps are not published. Instead, the birds are referred to as an 'individually colour-ringed population of Siberian jays that has been studied from 1989 onwards near Arvidsjaur'.[20]

20
Griesser, 'Do Warning Calls Boost Survival of Signal Recipients?', 4.

Combining the different field data allows me to create my own stories and images and relate to them from a sensorial perspective. By studying them with my design-informed ethnographic knowledge of the biological fieldwork, I can create my own path to follow the complexity of data collection and its tasks and challenges. The autographic data allow me to identify with the biologists, as in the case of Sinja and Lena, and notice differences in how they take notes, as different people collect data using differing practices, and with varying levels of experience and skill. At the same time, the material traces allow one to imagine the field environment and the conditions to which the biologists were exposed.

This second and third category of autographic data are usually disregarded during the process of digitalisation. In fact, these aspects will be filtered out entirely during the next step. Their existence ends in the field notebooks. Once the data are digitalised and homogenised, these differences become invisible, as does the opportunity to observe certain aspects of data collection. From my perspective, the datasets become *flat* with the elimination of the autographic data. However, for the biologists, only through this step – the literal flattening from three-dimensional notebooks to two-dimensional screens – do their data become manageable. They will be turned into formalised numbers and figures on an Excel sheet that represent the data plot.

I suggest a fourth, immaterial data category, which becomes visible only in the raw data, but is obvious in my ethnographic observations described in the previous chapters. The biologists' sensory, embodied, and implicit knowledge shapes and accompanies the research practices discussed in the previous chapter (Chapter 5: 5.3). This may be the embodiment of certain gestures, movements, or experiences that remain in the memories of the biologists, which means that certain field situations can remain in the biologists' memories more clearly than others. This knowledge is important during fieldwork because it accompanies *Situated Enskillment* (Chapter 4: 4.3), *Sensory Alignment* (Chapter 5: 5.3), *Situated Mediations* (Chapter 5: 5.4), and *Participant Behaviour Observation* (Chapter 5: 5.5). These practices are highly specific and require prior training in situ, familiarisation with the tools and methods, and alignment with the environment and birds. This knowledge is inevitable for data collection; however, when it comes to data analysis, it is filtered out. In short, from a biological perspective, this knowledge is lost, or it remains invisibly inscribed in the individual biologists' bodies.

6.1.2. Second Filter: Processing Raw Data into Analytical Data

A few weeks after data collection in Sweden, I return to the biologists' offices in Switzerland to discuss the next steps. After I make my way to the Anthropological Institute of the University of Zurich, where Michael and his project are based at that point, I enter the room at the end of the corridor, where I observe the field notebooks and hard drives lying on a desk. Seeing them immediately transfers me back to the field, where the very same books accompanied us every day to capture the biological and autographic data. Here, in the offices, they create a strange connection between the two working environments of the field and the office. For now, like the field, the biologists have to treat the data as they did the birds in the field, and, with the help of further filtering, separate the relevant and irrelevant information. They do this by transferring the raw data to digital files that replace the videos and notebook entries as containers for the filtrate. The resulting data and datasheets with the bodily measurements of the new birds will be transferred to a separate Excel sheet.

Here, in the offices, the field material is refined through the filtering process, as though it must first be made suitable for its new, clean environment. In the same way I cleaned the snow and soil from my boots before I entered the house in Sweden after a day in the field, the field data must be cleaned to rid it of all traces of the forest and the individual biologists before they enter the digital space of the university offices. In this moment, the operational space migrates from an analogue to a digital one. Now a biologist who may have never been to Arvidsjaur, nor seen a Siberian jay flying between the pines and birches in the boreal forests, sits in front of a computer with the field notebook, staring at a video recording of three birds around a feeder (Figure 56).

During fieldwork, the video camera worked as a filter, eliminating everything outside its frame. It transformed all auditory and visual information into digital data. The outcome is a high-resolution representation of the events that have been transformed into its two-dimensional equivalent; thus, the third dimension has been lost. The video camera works as a thick filter that omits everything visual outside the frame. At the same time, the value of this tool lies in its permeability. Thus, the video material has been only slightly filtered, and most of the visual information of the scene in which the biologists are interested is kept, whereas all other sensory experiences, except the auditory, are excluded. In fact, it is kept long-term and allows the biologists to transport the event from the field to the office, where they can replay the scenario while sitting on a chair behind a computer in a more comfortable environment than the field in Sweden. However, as raw data directly from the field, the videos do not tell the biologists much – yet. The files must first undergo an additional transformation process through further filtering.

A voice on the video comments:

> *OK, we are in Knott territory. The time is 09.54 and this is Kate, and the date is the 11th of March. There are three birds here, pink aluminium orange and red, pink aluminium, white, #, and yellow, yellow, orange, aluminium. [pause] Orange and red was just on the feeder [pause] that was yellow on the feeder, [pause] orange and red, [longer pause] double yellow again, [pause] and orange and red joins.*[21]

Kate's voice is interrupted by the rustling of her winter jacket as she moves her arms. Thus, the jacket also found its way into the field material as a testimony to the cold conditions, revealing the movement of her arms whenever she lifts the binoculars to study the colour IDs of the birds. This goes on for 15:02 minutes; however, I stop after 01:25 minutes and fast-forward through the video to obtain an overview. I have seen many other similar recordings, and it seems as though I will find nothing unusual in this one. For me, this is sufficient information, and I return to my text and continue writing.

However, the biologists transcribing the data need more patience. They must also study Kate's notebook entries to contextualise the data and note any unusual events that may have occurred that she may have written down, such as the presence of a raven, a predator of the jays. As Michael told me one day, 'The observational chart will in most cases not be transferred at all, it mainly serves for orientation and to get a feel for the territory'. The behavioural observation protocols in the field appear to work mostly as a method for (sensory) enskillment, as the behavioural observation protocols will be repeated in the offices based on the video recordings, thus discrediting those from the field as imperfect or incomplete.

21 Kate: Behavioural observation recording in *Knott* territory, 11/03/2015. File name: 00001 Kopie.MTS

While the relationship between the referent and its reference is symbolic in the behavioural observation protocol, the video recording still shows the birds as they appear in nature. Thus, it produces a translation of an event that occurred in the field. The videos are recorded while the behavioural observation protocols are conducted. Accordingly, when replaying them on computers in the offices, with the notebooks open to check for side-notes, a digital repetition of the event occurs – an on-screen re-enactment of what happened a few weeks earlier in the field. While many aspects are lost in the video, as it is a two-dimensional recording of one frame of a much denser event in nature, this replication of an otherwise ephemeral event also has its advantages. The biologists can pause, rewind, and fast-forward the videos to carefully analyse specific interactions. While the behavioural observation protocol took only 15 minutes in the field, its repetition in the office takes much longer. During this process, the biologists apply the same codes and attempt to identify the same behavioural patterns as they did during fieldwork (Figure 57).

This digital version of the events allows them to scrutinise the interactions between individual birds and then quantify them. The biologists can now further filter their observations from the video to the Excel sheet, thus repeating the entire fieldwork process one more time. However, the environment, tools, and methods, and necessary visual skills, shift considerably. The biologists are no longer exposed to the environment but sit in front of a computer on an ergonomic office chair. They can pause the video whenever necessary, whereas if they miss anything in the field, they lose data. Last, their operations are reduced to a digital environment; they no longer need binoculars, video cameras, or field notebooks. Through this process, the birds are geographically relocated. They are arranged on the data plot in a linear way, as opposed to their geographical location on the maps used for fieldwork.

During this process from video to Excel sheet, the biologists again extract only the relevant information. To filter the interactions of the birds, and to make them unambiguous and enable digitalisation, Kate must disappear from the video. Her voice will become silent and she will be replaced by numbers and figures. In addition to Kate, the trees, snow, landscape, feeder, branch, light, and birds themselves will be filtered out, and with them, the birds as living animals. In fact, the video recording as a file must disappear and be transformed into a new data environment that facilitates further operations. It becomes residue. Only a small percentage of *filtrate* of what can be seen in the video is kept and presented in the Excel data plots. Here, the biologists again combine digital operations, a framing interface, and their interest and skilled visions as situated mediations to fulfil this step. This step is closely related to the objectivisation of the individual biologists who are represented as neutral actors in the construction of scientific facts.

However, the territory names, dates and times, and the birds' IDs remain the same, allowing the biologists to merge the new data with their existing datasheets. With each step that brings the biologists closer to their output, they distance themselves more from the field and the birds, which are replaced with numerical representations. They apply filters to limit their focus even further to finally confirm their hypothesis and ensure that as few autographic data as possible make their way into the research results.

The video recordings and notebooks do not claim objectivity and are therefore much closer references to the field situation than the abstract Excel sheets. Their visual and haptic nature means they are directly bound to their origin in the field. They still carry aspects from the field, such as rainy weather that may have resulted in wrinkled pages or dead mosquitoes caught between pages. To echo Latour, the translation step is much smaller between the bird and the field notebook, where its behaviours are documented, than between the bird and a digital chart.[22] The field notebook offers a much richer representation of the field situation and, ultimately, the natural world. From a new materialism perspective, these field notes can even offer 'a glimpse into a parallel world of vibrant, powerful things'[23] as they have their own lives and make unexpected and unplanned aspects visible; they tell their own stories, as I have attempted to illustrate.

Last, these 'powerful things', 'vitality', and modalities that 'distribute agency'[24] become much more evident in the field notebooks than in the digital charts, where the various media are hidden. The supporting media thus do not play the role of neutral assistants, but the forms of knowledge and thought styles are greatly influenced by the employment of the mediators and skilled visions.[25]

When studying the correct page in Figure 58, biologist Jasmine did not even produce a written behavioural observation in *Fat Jana*. Instead, she wrote in capital letters, *'NO WRITTEN ONLY SPEAK TO THE CAMERA'*, either as a note to herself or to her colleague who will analyse the recording. Michael recently confirmed that fieldwork has been increasingly shifting towards digital recordings alone, with handwritten protocols being entirely omitted. These written protocols are mostly used during the first days of fieldwork when the biologists are getting used to observing the birds and practising aligning with them.

As Marine told me, this follows the thought style of technically-inclined biologists or those who base their research more on empirical observation than their virtuosity in the field. The technical approach requires different skills to

22
Latour, 'The More Manipulations the Better'.
23
Jane Bennett, 'Powers of the Hoard: Further Notes on Material Agency', in *Animal, Vegetable, Mineral: Ethics and Objects*, ed. Jeffrey Jerome Cohen (Washington: Oliphaunt Books, 2012), 239.
24
Ibid., 241.
25
Cf. Frederik S. Pötzsch and Bernt Schnettler, 'Visuelles Wissen', *Handbuch Wissenssoziologie und Wissensforschung*, 2007, 480.

those I have described thus far. When working with big datasets, the biologists require a broader skillset of informatics and programming to analyse and model their datasets. Both create different modes of knowledge production and work in different knowledge spaces and possibilities of thought: they create different knowledge. Accordingly, biologists must critically reflect on their tools and how they influence their perception and knowledge production process so that they can uncover blind spots and unconscious filtering processes, thereby taking responsibility for their knowledge production.

When the biologists filter data in their offices, they capture singular aspects that occurred in the forest of Arvidsjaur on digital datasheets, thereby extracting them from their location and obscuring the conditions of data collection. At some point, the time and date become mostly irrelevant and are condensed as timespans of several years: 'The study was conducted between 1990 and 2004….'[26] These timestamps shift slightly depending on which data the biologists are working with and publishing.

Along with this, the birds are removed from their environment and reterritorialised into digital data plots. In the field, coloured IDs are attached to the birds' legs. But in the datasets, they become detached from their physical context, existing only as combinations of numbers and letters aligned with territory names in the columns of Excel sheets. For as long as the research continues in the offices, the birds have become refined entries in these sheets, independent of the context and origin of the data. With this, the birds move as detached data points across the digital surfaces of the biologists' computers, ready to be processed further. Henceforth, their environment is no longer relevant. Their three-dimensional space has been replaced by a linear list. However, this list does not follow an alphabetical or chronological order but is organised

26 Magdalena Nystrand et al., 'Habitat-Specific Demography and Source-Sink Dynamics in a Population of Siberian Jays', *Journal of Animal Ecology* 79, no. 1 (2010): 267.

according to the actual locations of the territories in the field. In this way, the biologists have sought to translate the spatial logic of the map into a linear format, arranging the forest territories in the list so that those located near each other in the field also appear close together in the data.

Once the biologists and their assistants have spent several weeks transferring data to digital datasheets, an additional step of filtering has been fulfilled. In this way, the biologists transfer their visual material into a new graphical form through visual reduction and clarification. During data processing, the biologists filter out most of the research conditions, obscuring any aspects besides those that are relevant to their research. In this step, all ambiguity (from a natural scientific perspective) caused by the visual representations is eliminated, and what remains is clear, visual evidence. Only after this transformation can quantitative data analysis be undertaken. The more the data have been transformed by different filters, the more homogeneous they become. In the resulting chart, I cannot identify which data have been collected by whom or make any other assumptions about individual biologists and what they may have experienced in the field. Tracing their practices, as I did earlier (by means of the raw field material from Lena, Jasmin, and Sinja) is no longer possible.

6.1.3. Third Filter: Modelling Data into Results

Observation is a perfectly respectable concept in the natural sciences, but to make it part of scholarship, it ordinarily must be pulled into systems thinking in which ahistorical natural laws can be seen in action. Anna Tsing, an ecologist, explained to me that she loves fieldwork for its natural history observations, but when it comes time to publish an article, she must turn to theory, which, for her, refers to systems modelling, in which observations can be removed from their moments of contingent connection to speak to general, mechanical principles that sustain the world.[27]

Henceforth, the biologists also lock the notebooks and hard drives containing the video recordings away in archives, and their scientific project turns entirely into digital work, disregarding almost all visual operations of filtering that have led to this point. Now, they can start their actual data analysis and modelling to produce visualisations to confirm the hypothesis they developed prior to data collection. Only in very rare cases, if their results appear to be inconsistent, will they consult the field notebooks to check whether they missed anything or made a mistake, such as a typographical error, when transferring the data. This final step is shaped by a third layer of filtering that determines which data from the Excel sheet are evaluated and which data are disregarded.

Now the biologists follow the visual structuring principles that meet the codes of scientific representation. They follow the authority of schematic, *unambiguous* visual representation in the sciences, thereby naturalising the tools and practices employed by making them invisible. They use R[28] software to develop algorithms and write scripts of the raw data plots, which helps them confirm or reject their hypothesis by extracting and selecting data from the Excel sheet, coding it, and generating a visual result – a graph – as *proof*. In this step, the input data in their different stages become invisible, as do the algorithms that create the modelling. The resulting graphs appear to be a direct result of what the biologists encountered in the field, thus obscuring the many operational steps that led to the first data plots and served as input for their modelling.

27
Tsing, 'When the Things We Study Respond to Each Other', 239.
28
R is a statistical environment used by the biologists to analyse, model, and visualise statistical data. See https://www.r-project.org/.

The graphs obscure the fact that their outcomes are the result of a long operational process of bodily and sensory engagement of filtering practices. The final data charts or graphs published in papers are contextualised only by further information provided by the authors; reconstruction of the steps from the original source is impossible. All other traces – the fat, birds, biologists' handwriting, snow, and weather – have disappeared, as they are not relevant to understanding the results.

In summary, turning the birds into flat surfaces is a prerequisite for producing scientifically valid data. The available information must be filtered through several practices to achieve this transformation, which can be separated into three main steps of situated mediations: (6.1) filtering of raw data on the birds and capturing them on data carriers during fieldwork; (6.2) transformation of this raw field data into raw data plots based on digital operations that work as filters; and (6.3) modelling of the raw data plots with computer software until a *poor*ified version of data on the birds is available as the output. In all three steps, the biologists operate the filters. They are entangled with their practices and conduct the filtering. Thus, they have become inseparable from the apparatus of knowledge production. They are inscribed and situated within it rather than outside or even *above* it.

The process described in the previous section is almost like one of purification in a laboratory, or when applying filters with low permeability. What was initially a first-hand account of a single event observed in nature has now been transformed into a schematic representation that enables further data modelling. Through this step, the field and office have been separated in the sense that the field is hardly relevant anymore and the operational space reduced to data evaluation. New images will be developed based on these data plots, all of which are stored on hard drives and clouds, to *confirm* what, until now, has existed only as a hypothesis.

What was developed during several small steps that were not solely performed in a linear, consecutive order, but rather in a messy, multi-layered, entangled environment, cannot be retraced from the resulting graphs. In hindsight, the final images representing the outcome of the research in the coming papers appear to be one large step, produced directly from bird to paper; the life cycle of the data construction has been replaced by a brief explanation in the methodology section.

6.2. Relocating Birds and Biologists

Not only are the birds relocated from field to office during the filtering processes, but the biologists' working environment also changes. While the fieldwork is highly specific and dependent on a specific location – the boreal forests of Arvidsjaur – and cannot easily be replaced by a different location, the offices are less dependent on a particular geographical location. Instead, they depend on the biologists who analyse the data and thus have authority over them. During our collaboration, Michael and his team changed offices several times, depending on Michael's position and the relevant funding bodies.

When my fieldwork started, Michael's team had their offices in the Anthropological Institute of the University of Zurich. After the project's funding period had expired, Michael's office, along with his archives, was moved into his private basement. With the imposition of several COVID-19 lockdowns, the team's working environment moved into the digital space, with the biologists working remotely from their homes, which is where I would meet them. Since then, Michael has assumed a new position at the University of Konstanz. He, his research project, and, most importantly, the digital data (on hard drives and in archives) have thus, once again, moved offices and become embedded in a new working environment. Along with this, the size of the team and their respective roles also changed.

What remained the same throughout was the field in Sweden, except for the birds that have inevitably aged. Some (with or without IDs) may have died, and new nestlings may have been born and equipped with IDs. Some groups may have grown, some may have diminished, and others, as in the deforested territories, may have disappeared altogether. In this way, the territories may have changed slightly over the years, but all these events will have occurred in the boreal forests of Arvidsjaur. However, this sense of location experiences one last shift when the results are published as the final step of the research; the scientific results about the birds become entirely detached from a location once they circulate in scientific journals around the world, although they usually remain within a small group of scientific peers.

When studying the entire dataset, it becomes clear that the data are mainly produced in two environments: first, in the field and, second, in the office. While the first moves in a rather analogue and concrete space, the latter migrates almost entirely into the abstract digital space. Both fieldwork and office work are essential to the scientific study of the birds' behaviour and complement one another. These work environments are shaped by different practices, material, infrastructure, ways of thinking and knowing, skills, and sensory experiences. In the field, thinking is mainly shaped by data collection as thinking in practice; in the offices, thinking takes on a more analytical form, where data are interpreted, algorithms are written, hypotheses are confirmed/rejected, and new hypotheses defined. The biologists sit behind their computers, analyse data, program algorithms, exchange information with one another (on red sofas), and discuss the data and outcomes (behind adjustable desks). Once they have *confirmed* their hypothesis, they start writing and developing research papers, which they submit to scientific journals for publication.

The work in the field is shaped by pragmatism, prototyping, ingenuity, and creativity: from preparing the equipment for fieldwork to fixing the house and preparing food. However, the *doing* part, the fieldwork – observing the birds, and the tools and skills required for this, as well as the moments of despair on cold days when no birds appear – has disappeared from the results. Instead, this environment has now been replaced by human infrastructure provided by the institution, where facility managers take care of the equipment, technical staff fix the computers, and the biologists have lunch, and perhaps even dinner, in the canteen (where a great deal of thinking may also take place). It appears that once the biologists have proven their skills of surviving by themselves in the field, they can invest all their energy in their analytical process and output when they return to their offices, as the institution ensures that no other tasks distract them from their work, as long as they have sufficient funding and generate scientific output.

One additional aspect accompanies this process of relocation: during data collection, as I have described in Parts 1 and 2, the biologists immerse themselves in the field, not only regarding their research objective but also with regard to their entire bodies and sensorial attention. They experience the weather conditions, strenuous fieldwork, and the physical exertion of moving through the snow in the field. They also experience a *sensory alignment* (described in Chapter 5: 5.3) where they, to a certain extent, need to identify with the birds, imitate their calls, catch them, handle them – literally – and follow their ways of thinking when searching for their nests in summer. Thus, they engage with the birds on a multisensorial level, combining the auditory, visual, and haptic, as well as the 'in-between' of these senses.

Once fieldwork has been completed, these aspects are no longer relevant. The sensory engagement is now reduced to visual observations and sense-making on screens and in field notebooks, thus affirming my initial interest in visualisation. In hindsight, this initial interest is no surprise because it reflects what is publicly visible of scientific work. When the biologists analyse field videos in their offices, they no longer listen to the bird sounds as they would during fieldwork. They only (visually) observe the interactions of the birds and listen to the words of the biologist speaking on the video, who is interpreting the calls they hear during the observation or pointing out aspects that may not be visible on the video.

Therefore, the biologists' sensory attention is reduced to the visual during office work and thereafter. However, during fieldwork, the biologists are engaged with all their senses. They *observe* with their entire bodies. During these practices, they collect embodied and implicit knowledge that is (epistemologically) lost once their bodies are reduced to sitting in their office chairs and staring at computer screens. However, this knowledge remains inscribed on their individual bodies and becomes activated once they are exposed to the field again. To others, such as myself, this knowledge becomes perceptible when observing the different practices of sensory and bodily engagement in the field, as well as the different levels of experience and virtuosity with fieldwork. In this sense, the sensory and bodily engagement of the biologists during research only serves as a means to an end. Its capacity as a means for data collection is thus discredited and possibly also underestimated.

The contrast between fieldwork and office work is striking. Even though the fieldwork is crucial and forms the basis for the office work, until I started my research, the practices of the field were a mystery to me. I could imagine the processing of data on computers, but I had no concept of how the research team would obtain their data in the first place, how they would collect and store data, and how this would form the basis for a research paper.

Enriching vs Poorifying Data

Returning to my initial question, the gap between the research subject in the field and the object in the paper was one I could not bridge without conducting ethnographic research through my case study. This gap was not caused by a lack of information but rather by a lack of access. After the filtering process that transforms the birds into data, the life cycle of these data is eliminated. The practices, materials, and working conditions of scientific research are rarely publicly presented, as I have aimed to do.

This method of handling data is entirely different from practices in anthropology. Here, reflective and field diaries also involve the emotional and mental aspects of the biologists themselves; one can read about their fascination for their research and the activities that accompany data collection. An interview is relevant not only for the spoken aspects that can be turned into data but also for the unspoken, and the atmosphere and setting that embeds the interview situation within the social context encountered in the field. It appears to be the ethnographer's intention to reveal the relationships, entanglements, sensory engagement, and messiness of the field situation rather than to clean it up and present sterile data. As an ethnographer, I paid particular attention to what is filtered out in scientific research, as this is where I can observe the specificity of certain practices, such as sense-making, and often where the conditions of knowledge production become visible.

The scientific validity and acceptance of the data seem to be established by a sensory and atmospheric *surplus* that reveals the conditions of research and provides information about the research partners – the environments – and thus has the capacity to reveal social conditions. Conversely, what I filter *in* as part of a thick description in biology is filtered *out* for scientific validity. While Michael works in his office, purifying his empirical data and formulating his findings, I am sitting in my office, attempting to enrich my text with sensory descriptions of the field situations in which I found myself, along with Michael, in the spring of 2015 and 2020. I am not doing this as an author of a novel, but to allow my audience to engage with what I encountered during my fieldwork as precisely as possible, while being aware of my partial perspective, the only perspective I can have. With this, I attempt to add one piece to the puzzle of scientific work that usually remains a mystery to people outside the scientific community.

6.3. Thick Description by Means of Visualisation

I introduced the metaphor of filters to offer a new perspective on what Latour and Woolgar called 'cascades of inscription'.[29] The metaphor was used to understand the practices of knowledge generation in my case study differently. The concept of filtering is useful to attend to the practices, thus analysing the processes of data collection rather than independently analysing the results. However, this metaphor also has limitations and disadvantages. The concept of filtering is often closely related to refinement, purification, and essence, thus affirming the notion of improving and optimising the data. As I have attempted to argue, this process of filtering – and the *purified* results – while inevitable for knowledge production in evolutionary biology, is, from an STS perspective, closely related to a *poor*ification of data: the loss of complexity. One could also call it

29
Latour and Woolgar, *Laboratory Life*, 21.

a *thinning* of the data, with reference to the ethnographic method of thick description, as certain layers of information are filtered out.

Thus, filtering works as a productive metaphor only when viewing it *critically*. Otherwise, it affirms, in the natural sciences, an ongoing paradigm of objective data as data that appear neutral, that are not ambivalent, and represent clarity. However, I would like to emphasise that it is precisely the unfiltered, *dirty*, and *messy* data that make knowledge production in the natural sciences interesting. They allow one to complexify what has been simplified. For this reason, I conclude this chapter by suggesting the concept of a *multimodal thick description*, that is, a combination of the different (raw) materials that result from the process of knowledge generation among evolutionary biologists.

As I have illustrated in this chapter, data have different qualities when viewed from different epistemological perspectives. They enable different thoughts, tell different stories, and produce different insights. Therefore, I subjected this material as arte*fact*s of a process that aims to produce facts from an STS perspective and suggest a different way of viewing it. In so doing, I wish to address the effects of treating certain data as *waste products*. More precisely, those data are treated as a *residue* because materials that are filtered out do not *go to waste*, but are stored in the archives, a prerequisite for conducting my research. However, this biological residue does not become part of the analysis, and it is neither interpreted nor reflected on. It is filtered out.

By reintroducing these raw field materials into the discourse, I offer a dense presentation, a thick description, of the scientific knowledge production. In so doing, I also reveal the different modalities that accompany the scientific process. With this, I shift the perspective from *what* is thought to *how* the thinking occurs. This shift further allows transformation from a representational perspective to a performative one that focuses on the processes rather than the results. Rather than viewing the knowledge production in this study as a linear process from bird to data via visualisation, studying the overall dataset provides an understanding of the relationship between the different data materials as entangled networks. In them, the relationship between research objects, artefacts, and biologists becomes visible, as they form the basis of knowledge production, one of the most important aspects of making scientific sense of the world.

In ethnography, thick description is understood as a detailed account of the ethnographic observations that 'bring us [the audience] into touch with the lives of strangers',[30] 'what generality it contrives to achieve grows out of the delicacy of its distinctions, not the sweep of its abstractions'.[31] To attend to this 'delicacy of distinctions' requires considering what made the observed practices specific rather than generalising them into abstractions. Applying this to my case study means combining the datasets and the practices that lead to them. Rather than studying them from the *outside* and generalising them, thereby creating abstract representations, a thick description means the opposite: immersing oneself in the research conditions to study them from the inside and create a thick, detailed representation. Instead of filtering them out, my aim was to offer a *thick* account of the biologists' reality that, as declared in my introduction, brings epistemological and ontological invisibilities to light.

I conclude my discussion on the concept of a multimodal thick description by referring to the concept of infrastices, suggested by Ingold and Anusas: 'Western industrialized design produces objects by dividing surfaces from what we call "infrastices". By infrastices we mean all manner of

30
Clifford Geertz, *The Interpretation of Cultures* (New York: Basic Books, 1973), 16.
31
Ibid., 25.

electrical, chemical, and mechanical workings; their parts, structures, and conduits; and the energies, gases, and fluids they carry.' In a footnote, they add: 'We have coined the term "infrastitial" here (from infra = "below, beneath" + stare = "stand") as an alternative to "infrastructural", to avoid the latter's connotations of foundational support. From this, we derive the terms "infrastices" and "infrastitiality"'.[32]

While these authors are mainly referring to the hidden characteristics of objects beneath the surface, I would like to appropriate this term for my research in a more metaphorical way by suggesting that 'infrastices' are the production conditions (situated mediations) that lead to opaque surfaces; they may become invisible but are still inscribed on them.

I argue that the final graphs, even the raw, digital Excel data plots, lack infrastices. They are opaque surfaces that separate the sources of data – the birds – and their results. This final visual representation does not allow even a glimmer of the birds to shine through. It is entirely separated from the ontological space in which the data originated and has been moved to a new ontological and epistemological space that follows the logic of scientific rhetoric, image production, and representation. Accordingly, when taking the infrastices seriously and examining them, they can help change the status of scientific graphs and models as evidence towards the result of graphical operations with their origin in a research subject, the Siberian jays.

Visualisation is the interface between the sciences and the world, and a particularly important aspect of knowledge production. Currently the predominant instrument for presenting evidence, the visual way in which knowledge is transmitted influences how we think about the world. The scientific audience is usually confronted with the end of the life cycle of data, which, from my perspective

32
Tim Ingold and Mike Anusas, 'Designing Environmental Relations: From Opacity to Textility', *Design Issues* 29, no. 4 (2013): 58, https://doi.org/10.1162/DESI_a_00230.

as a designer and STS anthropologist, has become *dead* material, when compared with the raw material that led to it. This book can be understood as a multimodal thick description that addresses the different senses the biologists engage with during the research process, makes the biologists' standpoint clear, reveals the selective nature of knowledge production, illustrates the notion of thinking in practice, and situates the research.

Chapter 7: CONCLUSION

Data collection does not only occur through visual observation. Observation is a matter of sensory engagement – with eyes, ears, and hands – and it involves the body and listening as much as it does vision. My perspective on scientific knowledge production in the field sciences has shifted after five years of accompanying Michael and his team during fieldwork, online calls, and laboratory visits, attending to the sensory and bodily practices of their research process. By focusing on the boundaries of epistemologies and ontologies with which these field scientists engage during their work, my perception of the role of the visual shifted from it being *the* observational practice towards it being *part of* several observational practices, embedded in a wider sensory engagement. In my case study of evolutionary biologists, the visual was complemented by other sensory and bodily practices during data collection in the field. However, data collection is also guided by technologies and epistemologies that frame the act of observing. It was only during the processing and modelling of the raw data that the biologists' sensory engagement seemed to return primarily to the visual (i.e. the computer screen).

Bearing this process of filtering and transformation in mind, it is clear that the resulting '[i]mages or representations [that, i.e., the readers of scientific journals usually see] are not snapshots or depictions of what awaits us but rather condensations or traces of multiple practices of engagement'.[1] The final images are the result of a process of bodily and sensory practices, and not a simple mirroring of nature that 'awaits' the researchers in the field. What follows from this observation is that evolutionary biology appears to engage not only with *hard* facts based on visual evidence but also with several practices that could be rendered as *soft*,[2] as they require bodily and sensory knowledge that is hard to grasp and define. Thus, it cannot simply be formalised, but pertains to experience, practice, and virtuosity.

Given this shift from hard to soft, and the tension between the so-called *hard* (e.g. natural sciences) and *soft* (e.g. humanities) sciences, I summarise the most important insights, observations, and outcomes of my study that led to this observation. I do this regarding the natural scientific practices that, as I suggest, cannot be generalised as *soft* or *hard*, but are, instead, a process that moves between different stages of these conditions. I subsequently contextualise my observations in terms of their meaning for future development and disciplinary boundary-making, which enables me to draw on aspects that I excluded from my study and suggest topics for future research within the relevant fields. Finally, I conclude with an *Epilogue* in which I discuss a recent shift in the research practices of the evolutionary biologists I accompanied. This shift offers a new perspective on the practices of fieldwork and lays the foundation for new scientific endeavours in evolutionary biology.

1
Barad, *Meeting the Universe Halfway*, 53.

2
During one of our last interviews, Michael described his field as a rather *soft* science among the natural sciences. He argues that in physics, mathematics, and chemistry, *hard* knowledge is produced. This is because there is little doubt about certain facts, for instance, the existence of gravity. However, in the life sciences, such as evolutionary biology, the boundaries are more blurred, and the discipline is often more concerned with confirming (not proving) a hypothesis under the specific conditions of the research setting. This is also reflected in the research papers, where the possible conditions when the outcome can be considered as *true* must be discussed. Often, the storytelling plays a major role in convincing the audience (cf. Zoom interview with Michael, 24 May 2022).

Reflection on Previous Chapters

Adopting a design-informed approach in my case study provided different perspectives on the relevant scientific practices. It allowed me to attend to that which is usually invisible – the performative, processual, sensory, and embodied characteristics of the scientific process – which, as I have illustrated, usually disappear. Attending to these from both a design and an anthropological perspective allowed me to examine the scientific practices with a different focus, and redefine my understanding of design and anthropology. As suggested in the Introduction, the lens of a designer is a suitable starting point for bringing these different disciplines into the discourse. It allows one to focus on the sensory, emotional, and aesthetic aspects, thereby mediating between different ontologies and epistemologies. However, these aspects are not only limited to design discourse but are also popular in anthropology. Combining design and anthropology allows me to extend design into the field of anthropology and vice versa. Whereas anthropology provides methods and theories on the *cultural* and *social* aspects of specific practices, design focuses on the actual *practices* and *processes*. This contributes to a broader understanding of the scientific practices of my case study as cultural, social, embodied, and sensory, and instigates a discussion on the relationships between these disciplines, which seem to complement one another.

By focusing on design practices, I have brought anthropology and evolutionary biology into the discourse, despite the difference in research topics and thought styles. In both cases, design is employed as a recording practice to facilitate complexity and stabilise and transport observations. These formal-aesthetic design aspects are part of the *soft* aspects of the research processes, as they are defined by their sensory and bodily aspects, thinking in practice, and aesthetic engagement.

Despite this commonality in data-collection methodology, the role of design is reflected differently in both disciplines. In anthropology, *soft* aspects, such as the sensory and bodily immersion in the field and the visual representation thereof, are reflected as part of the methodology and formalised and conceptualised in the interpretation. In biology, these qualities are eventually filtered out to produce *hard* facts. In my case study, this sensory involvement, such as sensory alignment with the field and the Siberian jays, is performed only as a means to an end, for instance, to find and attract the birds. This difference is foremost the result of the quantitative and highly formalised approach, as opposed to a qualitative and sensory methodological approach. Again, these are the result of the historical development of the different disciplines that have resulted in almost opposing scientific worldmaking practices, thought styles, and meta-narratives that must be followed for the disciplines to be accepted in their respective scientific communities. Anthropology must meet the criteria of qualitative data collection and create evidence through consistency: by publishing primary data, using direct quotations, and including the role of the ethnographer in the reflections. These elements are often perceived as *objective* only in their raw state, along with the visible presence of the researcher in the results. In biology, it seems to be the other way around: only once the raw data have been processed and quantified, tidied of any subjective traces, transformed and refined, have *facts* been produced that are considered objective and robust.

Accordingly, whereas in ethnographic accounts, the audience can see first-hand visual material, such as photographs, the images in natural scientific journals are technical and formalised. Rarely can the audience of the latter see an image of the actual research subjects, such as the Siberian jays. The final images are usually small fragments extracted from the natural world, often representing highly specific aspects, and the result of a long process of *Preparing, Collecting, and Producing,* as I have illustrated.

During the research process, the evolutionary biologists, Siberian jays, boreal forests, and data-collection circumstances became increasingly invisible with each step. They developed from data collection as a situated practice, which is highly dependent on the biologists' involvement in the field and with the birds, to the processing of data that refer to the birds but are ontologically entirely detached from them and their environment. In this visual processing on computers, the biologists have also become somewhat arbitrary, as the processing and modelling can be performed by people who may have never been to Arvidsjaur.

During the practices of *Preparing* (Chapter 4), the biologists underwent what I conceptualised as situated enskillment, in which they were initiated into the fieldwork practices of the thought collective of the relevant researchers. This enskillment complemented the formal knowledge gained from their university training. As this knowledge was insufficient for fieldwork – a practice-based endeavour – this in situ enskillment was necessary. It encompassed practices of organising and preparing for fieldwork, of learning the reference system that ensures consistency in the data, and of navigating through the field and learning how to conceptualise the geography of the field. Most importantly, it involved learning how to work with the Siberian jays: how to identify and register them, and how to collect and document the relevant data.

During the practices of *Collecting* (Chapter 5), the biologists did not collect data with highly technical tools and media, as the technical appearance of the results suggest. Instead, they were engaged with their entire bodies and sensory perception in the field, particularly when attracting the birds in the different territories. They aligned themselves with their environment during wayfinding, and with the

Siberian jays when calling them. This alignment enabled them to find the birds and collect data from them. To ensure that the data collection was systematic and purposeful, and to frame their observations, the biologists used tools such as binoculars, protocols, and video recordings. By using these during fieldwork, the biologists became part of the observational apparatus that, only through the combination of *situated enskillment* and technical and epistemological *mediation,* enabled consistent data collection. I referred to this knowledge – of how to collect data in this specific study and account for relevant circumstances – as *situated mediations,* drawing on *situated enskillment,* embeddedness in discourse, and bodily and sensory practices.

After data have been collected in the field and transported to the office, the biologists, during *Producing* (Chapter 6), transformed, processed, and modelled those data, again based on situated mediations. The situated mediations in the office were also guided by specific practices, now mostly digital, which were driven by the researchers' hypothesis and epistemological interest in the data. Thus, they worked as *filtering* practices through a combination of technical tools and operations conducted by the biologists.

During this process, the traces that specified the field data and situated them in relation to the individual biologists and birds became invisible. In fact, it seems to have been the aim of the biologists to refine the data from that which situated them (e.g. autographic data such as handwriting that identified individual biologists). Through this filtering, the data became *objectified* in the sense that all their subjective, specific, and unique traces were eliminated. In this step, the biologists and the birds were not only relocated from field to office but also became less visible as *facts* began to emerge. As illustrated by Haraway's 'god trick', the biologists – except for their names on published papers – became neutralised and located somewhere outside the research. These steps of filtering were accompanied by a shift from a bodily and sensory engagement during the practices in the field, towards visual operations on computers during data processing.

In this final step, the implicit knowledge, involvement of the biologists with fieldwork, and their situated practices in Arvidsjaur have become redundant. The scientific papers affirm the traditional concept of objectivity as a visual construction of facts with a written account that meets the criteria of the standardised rhetoric, as they dominate the natural sciences, thus often rendering the information inaccessible to a wider audience. The result is a great leap from Siberian jays who live near Arvidsjaur to their reduced representation as evidence of a hypothesis in a scientific journal. They are presented as though the scientists have observed them from an outside perspective. However, as has become evident through my research, the biologists had to become entangled with the environment and the subjects they studied to collect data. From this perspective, they are situated within their research and not outside it. Thus, objectivity, in resonance with ecofeminist perspectives, must be understood as situated, as a partial perspective, and as something that must be conceptualised in relation to the practical engagements required by the field.

In this text, I have aimed to reveal the messiness of the environment – the natural world itself – in which the biologists are immersed with their entire bodies and senses, and from which the scientific *facts* stem. Rather than presenting clean, visually organised results of scientific work independent from the process of production, I have returned the natural world to the conversation by examining, complexifying, and entangling the Siberian jays, the evolutionary biologists, and the boreal forests, and revealing how they are intertwined. In this sense, I have attended to the issue of human entanglement and its situatedness within the natural world during data collection, which forms the basis for producing scientific facts.

In so doing, I have introduced two concepts that have allowed me to merge methods from both disciplines – anthropology and evolutionary biology – with the aim of overcoming the polarisation of the hard and soft sciences to reveal a wider range of conditions that may be neither hard nor soft. First, combining the ethnographic method of participant observation with the biological method of behavioural observation as *participant behavioural observation* allowed me to question the observational roles and instances that researchers from both fields assume. This concept helped to reveal that which is methodologically invisible – the gaps – in biology, by emphasising how the biologists engage with the natural world during observation. It allowed me to point out that evolutionary biologists should acknowledge the necessity of their temporary participation in the world of their research subjects to collect data. This could be an opportunity for them to reconsider their methodological approaches and engage with the public on a more accessible level.

Second, I introduced the concept of a *multimodal thick description* that borrows its meaning from the ethnographic concept of a thick description. However, I expanded it beyond text and included different modalities as they have become visible throughout my ethnographic project. I am aware of the limitations of publishing raw data in natural scientific journals. Nevertheless, I would like to call for more research in the field of STS and anthropology that reveals the modalities of scientific knowledge production, as in my case study, thereby bridging this gap. In particular, this could be done beyond the role of the visual, thus offering transparency and mediating between the (natural) sciences and the public.

Although I suggest that biology could learn from social anthropological methodology, I conclude by proposing that this learning is reciprocal. As Tsing has suggested, anthropologists could learn the method of *direct observation* from the natural sciences to address the challenges of the Anthropocene.[3] Direct observation, as the name implies, directly attends to the non-human world without humans (the biologists in my case study) as the mediators to explain *their* world to other humans. This could be one approach by which to avoid further anthropocentric worldviews in the sciences, given that they seem to be continuously perpetuated in both the *hard* and *soft* sciences. I did not manage to avoid this within the scope of this text; however, I did attempt to address it.

In light of the above, it was my aim to reintroduce the biologists, birds, and research conditions into the scientific discourse. By attending to the autographic data, I aimed to make the human and non-human actors of scientific knowledge production visible and offer a basis for reflection beyond the quantification and technical modelling of data. Rather than *flattening* the results, I have attempted to produce a multimodal thick description of the epistemologies and ontologies that enable scientific knowledge production. By enriching the data with the archival material that caught my attention as a designer, and by using an ethnographic methodology, I have attempted to overcome the gaps in the representation of scientific results by revealing the conditions and relationships that are part of the natural scientific undertaking of knowledge production.

In so doing, I have resituated the biologists as researchers and the Siberian jays not just as research subjects but as living animals within the study. I did this based on the observation that during the steps of *filtering*, the situatedness of the biologists also diminished along with a loss of complexity in the data. The individual biologists were visible

3
Tsing, 'When the Things We Study Respond to Each Other', 228.

in the raw data and could be identified as the authors of data. However, as the data were processed into a digital format, the authors increasingly disappeared, moving towards a generic scientific *we*. In addition to this, the data that first occupied physical and tangible spaces in material data carriers such as field notebooks, datasheets, blood and feather samples, and SD cards became more intangible ex situ through their transfer and filtration into Excel sheets. Once filtered, data can be shared via email (as Michael often did), multiplied on several hard drives as backups, transported on USB flash drives, and circulated among scientific peers. The scientific results circulate as papers in journals all around the world, becoming increasingly detached from their origin in the boreal forests.

Contextualisation and Outlook

These paradigms of scientifically representing – and thus also producing – worlds, must be reflected on and questioned to meet the challenges of the twenty-first century (climate crisis, mass extinction, and political and global instability), which are accompanied by a detached human–non-human relationship and a mistrust of the sciences. As plant ecologist and citizen of the Potawatomi Nation, Robin Wall Kimmerer,[4] states:

> [S]cientists are one of their [the stories of the natural world] translators and carry a larger responsibility for conveying their stories to the world. And yet scientists mostly convey these stories in a language that excludes readers. Conventions for efficiency and pre-

4 As distinguished Teaching Professor and Director of the Center for Native Peoples and the Environment at the State University of New York College of Environmental Science and Forestry, Robin Wall Kimmerer belongs to the thought collective of natural scientists. She is also a member of the Citizen Potawatomi Nation, a Native American people. She thus embodies the thought style, traditional knowledge, and Indigenous cultural practices of the Potawatomi. In this sense, she is a good example of bringing together two different kinds of worldmaking practices, namely scientific ones and spiritual-sensory ones, which reveal different human–nature relationships without pitting them against each other.

cision make reading scientific papers very difficult for the rest of the world and if the truth be known for us [scientists] as well.[5]

Given these urgencies, a shift is demanded in the relationships between the human and natural worlds, and, as Kimmerer states, scientists have the knowledge and ability to mediate these relationships. The technopositivist image of the natural sciences as a discipline that has *control* over nature based on technical manipulations no longer holds. This view must shift to one of the natural sciences having the ability to highlight entanglements, uncertainties, and complexities. It seems more necessary than ever to question the relationship between the human and non-human. However, to arrive at a more democratic relationship between these worlds, the narratives on nature must be conveyed in a more accessible way. This could be achieved through a transversal dialogue across disciplines where design and anthropology function as mediators to facilitate change.

One approach could be to reveal how scientists engage with the natural world prior to the production of facts. Attention to sensory and bodily engagement, implicit knowledge, visual skill, and situated training reveals the aspects that disappear during the research process. I have illustrated how, in the *softness* of evolutionary biology research, a different human–non-human relationship from that which is usually represented in scientific journals becomes visible. This relationship includes the sensory and the spiritual, as well as a responsibility for and fascination with the natural world. Attending to these allows for new narratives of the natural world. To echo Kimmerer: 'For what good is knowing unless it is coupled with caring?'[6]

Within the scope of this research, I have focused on the practices of knowledge production of *one* case study by examining the overall research process. The next step would be

5
Kimmerer, *Braiding Sweetgrass*, 345.
6
Ibid.

to conduct similar research with other cases to understand differences and similarities. Indeed, several sub-topics arose throughout my study that I suggest as topics for future research. Having embarked on my research endeavour from a design perspective, I began by focusing on the visual. However, throughout the process, the role of other senses became prominent. One aspect was the auditory alignment between biologists and birds, which I was able to study only to a limited extent within the scope of this project. Studying this aspect further could contribute to rearranging the hierarchy of the senses and re-evaluating other senses beyond the visual when discussing and developing scientific methods. While the auditory modality has recently entered the discourse in the social and cultural sciences, taking the auditory practices of the biologists more seriously could be one approach by which to broaden biological methods and arrive at new insights.

The role of forest management, which appears to be progressing further and shifting every year – affecting the study site and the behaviour of the birds – would be another topic worth investigating. Against the backdrop of the Anthropocene, I find the conflicting interests of the biologists (who aim to protect the natural world they are studying) and those eradicating it for profit, particularly interesting. The conflict is revealing of the research conditions and offers insights into global power relations and the politics underpinning the climate crisis. Given the changes to the study site associated with both forest management and climate change, it would also be important to further investigate how these developments have influenced the research and the questions posed by the biologists.

Within the scope of this project, I was able to reflect on and reveal my own practices as an ethnographer only partly. Nevertheless, I have still attempted to consider these aspects within the short 'patches', as I called them. It would be a worthy pursuit to attend more specifically to the sensory and bodily practices of data collection in ethnography, and to my own methods of data collection, developing a more in-depth account of them. Thus, I suggest developing more accounts in STS on the practices of knowledge production in the *soft* sciences.

Epilogue

I am in my office in Zurich in September 2021 on an online call with Michael and two of his PhD students: Camille, whom I met during my fieldwork, and Andrea, whom I have not met before. They are currently doing fieldwork in Sweden and Michael is joining the call from his home office in Switzerland. During our meeting, it suddenly strikes me that the conversation between the three biologists relates to what, in hindsight, could be viewed as a paradigm shift in their methodology.

The biologists are using new terms in their discussion of the fieldwork. They are discussing 'artificial intelligence', 'pipelines', and 'algorithms'. These are concepts that, until now, I had rarely heard from Michael. I struggle to follow their conversation, which makes me realise that this group of evolutionary biologists are undergoing a transformation prompted by AI, not only in their analytical practices in their offices but also in their fieldwork, situated mediations, and filtering practices. What had, until now, fascinated me from an STS perspective – the remarkable resistance to using digital technologies in data collection and processing, which led to the extensive use of seemingly anachronistic practices in this evolutionary biology study – has suddenly changed.

Henceforth, the main task during fieldwork will be to produce video recordings of the birds' interactions in a manner that allows analysis not through human filtering but by AI. To achieve this, the video recordings must be produced in a more formalised way and human observations must become secondary. The biologists will have to follow a strict protocol for recording events in the field so that the AI can filter them, produce datasets, and further quantify the birds. In short, the AI will replace the biologists during the second layer of filtering, while the way in which the data are produced in the first layer will be geared towards the AI's ability to process them. The biologists' main task will be to develop and apply the correct code to the data plots to model them.

The shift I refer to here is one from human labour, sensory engagement, and intelligence towards the automation of data collection, and particularly filtering. While, until now, the filtering processes were conducted by the biologists themselves, in conjunction with certain tools, these processes will now be taken over by technology, further obscuring them as they disappear behind an algorithm. Once their AI has been sufficiently trained, the biologists would not even have to view the videos themselves, as they will be directly translated into datasets by the AI. In addition, the only witnesses to the events that will be transformed into data will be the biologists who made the video recordings in the field.

From the biologists' perspective, this represents substantial progress in their data processing and their research in general. Soon they will be able to process and analyse a much larger dataset, address different research questions, and increase their research output. They will further quantify their field data and, among their scientific peers, their data will appear even more robust. Now, AI can even analyse previous video material collected over the years. However, this will be possible only if the video files are suitable and the AI has been trained to analyse them, a process that could take months or even years. On this video call, Michael mentions that his research group will be the first field biologists to work with AI data analysis. Until now, this practice has been performed only in laboratories under controlled conditions.

Now I consider it even more important than ever to attend to the ways in which knowledge has been constructed in this study. By focusing on the practices from a sensory and bodily perspective, I hope to have provided more transparency about the conditions of scientific knowledge production. However, having emphasised their role in knowledge production, the biologists should reflect on this when these practices are replaced and their scientific work is digitalised. There are informal aspects that may not directly influence the data itself but do affect the data-collection process and the fieldwork, and thus should not be overlooked or replaced. One such aspect is the behavioural observation protocol, an important tool for training vision. The general sensory engagement with the field is also important for making observations in the first place and, based on empirical observations, developing new questions.

Lastly, as Michael himself emphasised recently, the role of humour and joy during fieldwork should not be underestimated. As this component of data collection can become partly redundant and potentially boring, the biologists' attention and capacity to focus may be affected. From this perspective, it is important to keep the biologists busy during fieldwork and maintain their attention through focusing media, such as notebooks and protocols that frame observations and train the eye. However, there is another aspect to the joy of doing fieldwork: it attaches the researchers to the field (i.e. *nature*). They gain explicit and implicit knowledge about the natural world, learn about processes, entanglements and dependencies, and eventually understand how life on earth is interdependent and why humans are dependent on ecosystem services.[1] Fieldwork might be one reason why scientists become activists, standing up for the conservation and protection

1 Emilie Crouzat et al., 'Researchers Must Be Aware of Their Roles at the Interface of Ecosystem Services Science and Policy', *Ambio* 47, no. 1 (1 February 2018): 97, https://doi.org/10.1007/S13280-017-0939-1.

of nature. Scientists, often from societies with higher CO_2 emissions, constitute a small group of reliable protectors of nature, alongside, for example, Indigenous societies. Scientists' voices are particularly important when it comes to defending global change because their activism stems from a deep and substantial understanding of nature and its entanglements. I suggest that the depth of understanding derives from seeing challenges associated with climate change, biodiversity loss or ecosystem shifts in their results, and from being emotionally attached to the field. I have heard natural scientists speaking about nature as 'their one god', 'the only thing that brings them ease', or 'the place where they find spirituality'.

If fieldwork is reduced or stopped because the biologists have made themselves redundant, CO_2 emissions might be slightly reduced, although big data and AI also have significant CO_2 footprints. However, stopping fieldwork may come at a price: the attachment to nature may change to detachment. Scientists might lose their oversight of the importance of protecting the environment, thus contributing to, rather than defeating, the acceleration of global change.[2] In addition, this detachment might not just be from nature but also from other social beings such as fellow researchers because, as described earlier, fieldwork occurs in highly complex social settings, different from those of office work. If field scientists turn into informaticians, they may become more and more detached from their social surroundings and turn their attention towards big data sets. Big data has implications for the biologists and their practices. While, until now, the skills and education of field biologists were required for this research project, henceforth, the ability to handle big data, programming skills, and literacy with digital tools will be required; thus, scholars with an education in bioinformatics will be needed. Biologists are

2
Masashi Soga and Kevin J. Gaston, 'Extinction of Experience among Ecologists', *Trends in Ecology & Evolution* 40, no. 3 (March 2025): 212–15, https://doi.org/10.1016/j.tree.2024.12.010.

no longer entering the field. They do not need to be able to identify the Siberian jays, register them, observe them, or document their behaviour. They do not need to *prepare* and *collect*. They must only *produce* by programming an algorithm and training the AI. The focus of the research project is thus no longer on the collection of data, and more resources are being invested in processing these raw data and obtaining as much information as possible.

However, speaking again in 2025 to Michael and the co-PI, Miya, I gain an additional perspective. Having thought about AI in the Siberian Jay Project over the past couple of years, Michael and Miya agree that AI and technologisation, in general, lead to what they call a disconnect. They use 'disconnect' to mean a lost relationship to the field, the birds, forests, and the mud, eventually to the natural world. To them, even after the introduction of AI, fieldwork still means 'to get your hands dirty, to expose yourself to nature, and to be out there.' While the high-resolution technical data are helpful for their research, they still consider it vital to maintain a social-emotional relationship with what is in front of their eyes: 'reality', as Michael calls it. Otherwise, *we*, as a society, run the risk of losing touch with the real world, which must be avoided given the problems of our times.

Returning to the quotation by Haraway in Chapter 1, the matters, stories, knots, thoughts, descriptions, and ties of knowledge production will change with this technological development. New questions will arise in relation to how knowledge is produced, what kind of skills and tasks can be delegated to an AI, and how the meaning of the field is shifting. This text lays the foundation for addressing questions of big data, the digitalisation of research, and the shifting role of biologists themselves. These topics are not part of this project, but I consider them worth studying in the future, as they once again shift the epistemologies, ontologies, and ethics of knowing.

Territory Names

Angel (Reivo)
When Julian and his partner, Barbara, were looking for birds in 2014, they encountered two Sámi men hunting. Barbara was struck by the angel-like beauty of one of them. As the biologists had found a new group of birds there and a new name was required, Barbara suggested calling it *Angel.* In this way, she – and the situation that led to the name – became inscribed into the study site.

Akkavare (Managed)
Akkavare is a geographical reference to a hill near the territory.

Baggins (Reivo)
Michael and Jan were expanding their study site to include *Reivo* in 2000. Baggins, the name of a Hobbit family in the novel *The Lord of the Rings* by J. R. R. Tolkien was chosen, as it was next to the territory *'Lord'*, also named after the book. The two territories used to be closer together, but, in 2022, *Baggins* had moved some way away from *Lord* owing to the movement of the birds.

Basilika (Reivo)
This territory was named by Magdalena, one of three PhD students together with Michael and Sönke, who were part of the 'Siberian Jay Project' when Jan was still the PI. Contrary to her peers' suggestions, Magdalena did not want her first name inscribed into the study site. Instead, she suggested calling it *Basilika* as she enjoyed basil on her pizza and found the name fitting.

Blot (Managed)
Blot is another territory that shifted its location. The name refers to an incident Michael had during fieldwork in spring 2000 when he caught birds there for the first time. While he was trying to set up the net, he cut himself with a knife when cutting down a branch. As the cut was deep and quite messy with a lot of blood, Michael glued the wound on his finger with superglue, technically the same as medical glue although it is not sterile and is normally used for radio tags. He fixed it with additional tape and was able to continue working in what then became the territory *Blot.*

Fat Jana (Fat Road)
When Kate and I were doing fieldwork in *Fat Road* in 2015, we had set up a feeder and installed a camera. Kate began to conduct the behaviour observation protocol and the video recording with the oral descriptions. The birds were pulling quite firmly on the feeder until it suddenly fell down. I decided to help but, at this time, I was still very wobbly on the skis. I fell several times in front of the camera, while trying to get to the fat that was only a few meters in front of me. This (recorded) incident caused great amusement for everyone and resulted in the name *Fat Jana.*

Fat Moose (Fat Road)
When the field station of the Siberian Jay Project was newly based in Auktsjaur, once Michael became the PI, the biologists passed the road and placed some fat there, giving rise to the name *Fat Road.* In autumn 2013, Jonathan, a long-term field assistant and friend of Michael who had done his PhD with Jan, saw a large moose passing the road there.

Fello (Reivo)
Fello is a Sámi word to describe a small house that has been taken down and no longer stands where it used to. There are many 'fellos' in this area.

Fyra (Reivo)
Fyra is the Swedish word for four. It was chosen because the biologists had found four birds belonging to the same group, when they caught birds for the first time in this territory.

Glottje (Managed)
Relates to the nearby village, named *Glottje*.

G-liden or Guortesliden (Managed)
Relates to a geographical location, a hill nearby. The shortform *G-liden* is used as an abbreviation for the GPS system, as it is otherwise too long.

G-stjärn or Guortestjärn (Managed)
Refers to a geographical location to the north of a small lake nearby. The shortform *G-stjärn* is an abbreviation for the GPS system.

Guorbavagge (Reivo)
Guorbaive is a hill near where Jan caught birds in 2001. As there is a small, beautiful mossy valley nearby, Jan decided to change the name slightly and called it *Guorbavagge*.

Kara (Managed)
Kara was the name of the assistant in 1999.

Knott (Managed)
Knott are the small midges that were, according to Michael, 'literally everywhere' in summer 2000 when he caught this group for the first time. There numbers were so high that Michael decided to ring the birds in the car because 'the knotts crawl into your ears and nose and everywhere. It's horrible'.

Laxtjärn (Reivo)
Relates to the lake nearby. Michael assumes that there probably used to be salmon in the lake. However, this would be the origin of the name of the lake, not of the territory.

Lord (of the ring) (Reivo)
In the year that Jan and Michael wanted to extend the study site in *Reivo*, they put up fat everywhere to see if birds would come. And then, as described by Michael, 'Jan and I caught seven to eight birds at once. They were from different groups though, and some were kids who disappeared again. We thought: so many rings, it's like *Lord of the Rings*. Therefore, to make it easy we just called it *Lord* and skipped the rings'. After that a new territory was found next to *Lord*, which was then called *Baggins*, which also refers to *The Lord of the Rings*.

Maderängen (Reivo)
Ängen means grassland in Swedish and *Mader* is the name of a farm nearby. The biologists decided to merge the two names into *Maderängen*.

Mader (Reivo)
Refers to a farm nearby.

Måskapie (Managed)
Near Måskomyran and relates to the geographical area around the surrounding territory.

Måskomyran (Managed)
A geographical reference.

Mabuse (Reivo)
Dr Mabuse is the main character in a book by Norbert Jacques from 1920, which has been turned into a film by Fritz Lang (another name given by Jan).

Metro (Reivo)
This is also a name given by Jan, who likes films. This name refers to Fritz Lang's silent film, *Metropolis*, from 1927.

Nadine (Managed)
Nadine was the name of the field assistant in 1998.

Norna (Reivo)
Jan gave this name, which relates to an orchid that grows here.

Rågången (Managed)
Swedish word for a rough bordermarking in the forest or between farmlands, with a blurry boundary of about 5–10 metres. This border line is called *rågång*, and one is located near this territory.

Södra G(uortes) (Managed)
Geographical name relating to a territory to the south of a small lake nearby.

Spång (Reivo)
Paths that are built from wood across the swampland are called *Spång*.

Take 5 (Reivo)
This is similar to *Fyra* and was established in the same year. It describes how the biologists had five birds in the net when they were catching for ringing and measuring.

Trea (Reivo)
Like *Take 5* and *Fyra*, but there were only three birds in the net.

Troll (Managed)
In 2011, Michael and his team returned to this territory for the first time in a few years, and they found birds. There was one bird among them that they had caught before in *Kara* in 2007. The bird had been missing since then. In Swedish, 'trolla' is used for something that is difficult, that can perform witchcraft, or that can do magic tricks. *Troll* was an apt name for this bird that had been 'invisible' for a while.

References

Allen, Gary. 'Spatial Abilities, Cognitive Maps, and Wayfinding: Bases for Individual Differences in Spatial Cognition and Behavior'. In *Wayfinding Behavior: Cognitive Mapping and Other Spatial Processes,* edited by Reginald George Golledge, 46–80. Baltimore: Johns Hopkins University Press, 1999.

Anderson, Gemma. 'On Drawing as a Way of Knowing'. In *Drawing as a Way of Knowing in Art and Science,* 3–27. Bristol: Intellect, 2017.

Barad, Karen. *Meeting the Universe Halfway: Quantum Physics and the Entanglement of Matter and Meaning*. Durham, NC: Duke University Press, 2007.

———. 'Diffracting Diffraction: Cutting Together-Apart'. *Parallax* 20, no. 3 (2014): 168–87. https://doi.org/10.1080/13534645.2014.927623.

Bennett, Jane. 'Powers of the Hoard: Further Notes on Material Agency'. In *Animal, Vegetable, Mineral: Ethics and Objects,* edited by Jeffrey Jerome Cohen, 237–69. Washington, DC: Oliphaunt Books, 2012.

Berger, David Nathaniel, Nikita Bulanin, Lola García-Alix, Marianne Wiben Jensen, Signe Leth, Ena Alvarado, Dwayne Mamo, et al. '2 IWGIA – The Indigenous World – 2024'. Copenhagen: International Work Group for Indigenous Affairs (IWGIA), 2024.

Bergman, Ingela, and Lars Östlund. 'A Sacred Tree in the Boreal Forest: A Narrative about a Sámi Shaman, Her Tree, and the Forest Landscape'. *Human Ecology* 50, no. 6 (2022): 1023–33. https://doi.org/10.1007/s10745-022-00365-x.

Causey, Andrew. *Drawn to See: Drawings as an Ethnographic Method*. Toronto: University of Toronto Press, 2017.

Charmaz, Kathy. *Constructing Grounded Theory: A Practical Guide through Qualitative Analysis*. London: SAGE, 2006.

Coccia, Emanuele. *The Life of Plants: A Metaphysics of Mixture*. Cambridge: Polity, 2019.

Cross, Nigel. 'Design Research: A Disciplined Conversation'. *Design Issues* 15, no. 2 (1999): 5–10.

———. 'Designerly Ways of Knowing: Design Discipline Versus Design Science'. *Design Issues* 17, no. 3 (2001): 49–55. https://doi.org/10.1162/074793601750357196.

Crouzat, Emilie, Isabelle Arpin, Lucas Brunet, Matthew J. Colloff, Francis Turkelboom, and Sandra Lavorel. 'Researchers Must Be Aware of Their Roles at the Interface of Ecosystem Services Science and Policy'. *Ambio* 47, no. 1 (February 1, 2018): 97. https://doi.org/10.1007/S13280-017-0939-1.

Cunha, Filipe C. R., and Michael Griesser. 'Who Do You Trust? Wild Birds Use Social Knowledge to Avoid Being Deceived'. *Science Advances* 7, no. 22 (2021): 1–6. https://doi.org/10.1126/sciadv.aba2862.

Dance, Amber. 'Art Graft: Putting an "A" into "STEMM"'. *Nature* 590 (2021): 351–53.

Daston, Lorraine, and Peter Galison. 'The Image of Objectivity'. *Representations* 40, no. 1 (1992): 81–128.

———. *Objectivity*. New York: Zone Books, 2010.

Despret, Vinciane. *What Would Animals Say If We Asked the Right Questions?* Minneapolis: University of Minnesota Press, 2016.

Diefenthaler, Annette. 'Problemsolving'. In *Design Dictionary: Perspectives on Design Terminology,* edited by Michael Erlhoff and Tim Marshall, 307–8. Basel: Birkhäuser, 2007.

Dippel, Anne. 'The Big Data Game: On the Ludic Constitution of the Collaborative Production of Knowledge in High-Energy Physics at CERN'.

Naturwissenschaften, Technik und Medizin 25, no. 4 (2017): 485–517.
Dobler, Judith. 'Collaborative Imaging: The Communicative Practice of Hand Sketching in Experimental Physics'. In *Proceedings of DRS 2016: Future – Focused Thinking, Vol. 3, Sec. 5 Aesthetics, Cosmopolitics and Design,* 997–1011. Design Research Society, 2016.
———. 'Drawing Together: Collaborative Design Practices in Experimental Physics'. In *Nordes: Design + Power.* Nordes, 2017.
———. '"Flatten the Curve": The Performative Embodiment of Image, Speech, and Gestures in Scientific Communication from a Design Research Perspective'. In *NERD – New Experimental Research in Design 2,* 24–35. Basel: Birkhäuser, 2021. https://www.degruyterbrill.com/document/doi/10.1515/9783035623666-003/html.
———. 'Collaborative Drawing as Knowledge Practice'. In *Approaches to Drawing in Architectural and Urban Design,* edited by F. Colonnese, N. Grancho, and R. Schaeverbeke, 489–504. Newcastle upon Tyne: Cambridge Scholars Publishing, 2024.
Drucker, Johanna. 'Graphesis: Visual Knowledge Production and Representation'. *Poetess Archive Journal* 2, no. 1 (2011). https://paj-ojs-tamu.tdl.org/paj/article/view/4.
———. *Graphesis: Visual Forms of Knowledge Production.* Cambridge, MA: Harvard University Press, 2014.
Fleck, Ludwik. *Genesis and Development of a Scientific Fact.* Chicago: University of Chicago Press, 1979.
Flick, Uwe. *Qualitative Sozialforschung: Eine Einführung.* Reinbek bei Hamburg: Rowohlt Taschenbuch Verlag, 2017.
Frank, Adam, Marcelo Gleiser, and Evan Thompson. *The Blind Spot: Why Science Cannot Ignore Human Experience.* Cambridge, MA: The MIT Press, 2024.
Frayling, Christopher. 'Research in Art and Design'. *Royal College of Art Research Papers* 1, no. 1 (1993): 1–5.
Frow, Emma K. 'Drawing a Line: Setting Guidelines for Digital Image Processing in Scientific Journal Articles'. *Social Studies of Science* 42, no. 3 (2012): 369–92.
Galison, Peter. 'Objectivity Is Romantic'. In *Humanities and the Sciences,* edited by Jerome Friedman, Peter Galison, and Susan Haack, 15–43. New York: ACLS, 2000.
Gansterer, Nicolas. *Drawing a Hypothesis: Figures of Thought.* Edited by Gerald Bast. Vienna/New York: Springer, 2011.
Geertz, Clifford. *The Interpretation of Cultures.* New York: Basic Books, 1973.
Geismar, Haidy. 'Drawing It Out'. *Visual Anthropology Review* 30, no. 2 (2014): 97–113.
Gell, Alfred. 'How to Read a Map: Remarks on the Practical Logic of Navigation'. *Man* 20, no. 2 (1985): 271–86.
Goodman, Nelson. *Ways of Worldmaking.* Brighton: Harvester Press, 1978.
Goodwin, Charles. 'The Blackness of Black: Color Categories as Situated Practice'. In *Discourse, Tools and Reasoning: Essays on Situated Cognition,* edited by Lauren B. Resnick, Roger Säljö, Clotilde Pontecorvo, and Barbara Burge, 111–40. New York: Springer, 1997.
Gould, John. *The Birds of Australia.* London: Richard and John E. Taylor, 1848.
Grasseni, Cristina. 'Skilled Vision: An Apprenticeship in Breeding Aesthetics'. *Social Anthropology* 12, no. 1 (2004): 41–55.
———. 'Skilled Visions: Toward an Ecology of Visual Inscriptions'. In *Made to Be Seen: Perspectives on the History of Visual Anthropology,* edited by Marcus Banks and Jay Ruby, 19–44. Chicago: University of Chicago Press, 2007.
———. *Skilled Visions: Between Apprenticeship and Standards.* Edited by Cristina Grasseni. New York; Oxford: Berghahn Books, 2009.
———. 'Conference on Neuroesthetics – Seeing Knowing: Vision, Knowledge, Cognition, and Aesthetics'. In *Skilled Visions: Ecologies of Belonging and Sensorial Apprenticeship.* School of Optometry and Vision Science Program, University of California, 2014.
———. 'Beauty as Skill and "Common Sensing"'. In *Anthropology and Beauty: From Aesthetics to Creativity,* edited by Stephanie Bunn. London: Routledge, Taylor & Francis Group, 2018.
Grasseni, Cristina, and Thorsten Gieser. 'Introduction: Skilled Mediations'. *Social Anthropology* 27, no. 1 (2019): 6–16.
Griesser, Michael. 'Referential Calls Signal Predator Behavior in a Group-Living Bird Species'. *Current Biology* 18, no. 1 (2008): 69–73. https://doi.org/10.1016/j.cub.2007.11.069.

———. 'Do Warning Calls Boost Survival of Signal Recipients? Evidence from a Field Experiment in a Group-Living Bird Species'. *Frontiers in Zoology* 10, no. 1 (2013). https://doi.org/10.1186/1742-9994-10-49.

Griesser, Michael, and Jan Ekman. 'Nepotistic Alarm Calling in the Siberian Jay, *Perisoreus Infaustus*'. *Animal Behaviour* 67, no. 5 (2004): 933–39. https://doi.org/10.1016/j.anbehav.2003.09.005.

Griesser, Michael, Peter Halvarsson, Szymon M. Drobniak, and Carles Vilà. 'Fine-Scale Kin Recognition in the Absence of Social Familiarity in the Siberian Jay, a Monogamous Bird Species'. *Molecular Ecology* 24, no. 22 (2015): 5726–38. https://doi.org/10.1111/mec.13420.

Griesser, Michael, and Sarah Lagerberg. 'Long-Term Effects of Forest Management on Territory Occupancy and Breeding Success of an Open-Nesting Boreal Bird Species, the Siberian Jay'. *Forest Ecology and Management,* no. 271 (2012): 58–64.

Grimshaw, Anna, and Amanda Ravetz. 'Drawing with a Camera? Ethnographic Film and Transformative Anthropology'. *Journal of the Royal Anthropological Institute* 21, no. 2 (2015): 255–75.

Guggenheim, Michael. 'The Media of Sociology: Tight or Loose Translations?' *British Journal of Sociology* 66, no. 2 (2015): 345–72.

Hadfield, Michael G., and Donna J. Haraway. 'The Tree Snail Manifesto'. *Current Anthropology* 60, no. 20 (2019): S209–35.

Haraway, Donna. 'Situated Knowledges: The Science Question in Feminism and the Privilege of Partial Perspective'. *Feminist Studies* 14, no. 3 (1988): 575–99.

———. 'The Promises of Monsters: A Regenerative Politics for Inappropriate/d Others'. In *The Haraway Reader,* 63–124. New York: Routledge, 1992.

———. *When Species Meet.* Minneapolis: University of Minnesota Press, 2008.

———. *Staying with the Trouble: Making Kin in the Chthulucene.* Durham, NC: Duke University Press, 2016.

Harding, Wendy. 'Anna Tsing, Heather Swanson, Elaine Gan, Nils Bubandt (Eds.), Arts of Living on a Damaged Planet: Ghosts and Monsters of the Anthropocene'. *Miranda,* no. 16 (2018): 0–5. https://doi.org/10.4000/miranda.11648.

Hoffmann, Christoph, and Alexandre Métraux. 'Working with Instruments: Ernst Mach as Material Epistemologist, a Short Introduction'. *Science in Context* 29, no. 4 (2016): 429–33.

Hoffmann, Christoph, and Barbara Wittmann. 'Introduction: Knowledge in the Making: Drawing and Writing as Research Techniques'. *Science in Context* 26, no. 2 (2013): 203–13.

Howes, David. *Sensual Relations: Engaging the Senses in Culture and Social Theory.* Ann Arbor: University of Michigan Press, 2003.

Hugentobler, Hans Kaspar, Claudia Mareis, Franziska Nyffenegger, Ulrike Reichhardt, and Philip Zerweck. *Designwissenschaft und Designforschung: Ein Einführender Überblick.* Luzern: Hochschule Luzern, 2010.

Ingold, Tim. 'The Temporality of the Landscape'. *World Archaeology* 25, no. 2 (1993): 152–74.

———. *Lines: A Brief History.* New York: Routledge, 2007.

———. *Making: Anthropology, Archaeology, Art and Architecture.* New York: Routledge, 2013.

———. *The Life of Lines.* New York: Routledge, 2015.

———. *The Perception of the Environment: Essays on Livelihood, Dwelling and Skill.* New York: Routledge, 2000.

Ingold, Tim, and Mike Anusas. 'Designing Environmental Relations: From Opacity to Textility'. *Design Issues* 29, no. 4 (2013): 58–69. https://doi.org/10.1162/DESI_a_00230.

Istomin, Kirill V., and Mark J. Dwyer. 'Finding the Way: A Critical Discussion of Anthropological Theories of Human Spatial Orientation with Reference to Reindeer Herders of Northeastern Europe and Western Siberia'. *Current Anthropology* 50, no. 1 (2009): 29–49.

Jorgensen, Danny L. *Participant Observation: A Methodology for Human Studies.* Thousand Oaks, CA: Sage Publications, 1989.

Kammer, Stephan, Markus Klammer, Armin Schäfer, Richard Shiff, Jutta Voorhoeve, and Barbara Wittmann, editors. *Spuren Erzeugen: Zeichnen und Schreiben als Verfahren der Selbstaufzeichnung.* Zürich/Berlin: diaphanes, 2009.

Kappeler, Peter M. *Verhaltensbiologie.* Göttingen: Springer Spektrum, 2020.

Kimmerer, Robin Wall. *Braiding Sweetgrass: Indigenous Wisdom, Scientific Knowledge and the Teachings of Plants*. Minneapolis: Milkweed Editions, 2013.

Knorr-Cetina, Karin. *Epistemic Cultures: How the Sciences Make Knowledge*. Cambridge, MA: Harvard University Press, 1999.

Kohn, Eduardo. *How Forests Think: Toward an Anthropology Beyond the Human*. Los Angeles: University of California Press, 2013.

Krauthausen, Karin. 'Vom Nutzen des Notierens, Verfahren des Entwurfs.' In *Notieren, Skizzieren. Schreiben und Zeichnen als Verfahren des Entwurfs,* edited by Karin Krauthausen and Omar W. Nasim, 7–26. Zürich/Berlin: diaphenes, 2010.

Krauthausen, Karin, Stephan Kammer, Rüdiger Campe, Omar W. Nasim, and Ralph Ubl, editors. *Notieren, Skizzieren, Schreiben und Zeichnen als Verfahren des Entwurfs*. Zürich/ Berlin: diaphanes, 2010.

Latour, Bruno. 'Visualisation and Cognition: Drawing Things Together'. In *Knowledge and Society: Studies in the Sociology of Culture Past and Present: A Research Annual, Vol. 5, 1984,* edited by Henrika Kuklich and Elizabeth Long, 1–40. Greenwich, CT: JAI Press, 1984.

——. 'Circulating Reference: Sampling the Soil in the Amazon Forest'. In *Pandora's Hope: Essays on the Reality of Science Studies,* 24–79. Cambridge, MA: Harvard University Press, 1999.

——. 'The "Topofil" of Boa Vista: A Photo-Philosophical Montage'. In *Pandora's Hope: Essays on the Reality of Science Studies,* edited by Bruno Latour, 24–79. Cambridge, MA: Harvard University Press, 1999.

——. 'Drawing Things Together: Die Macht der Unveränderlich Mobilen Elemente'. In *ANThology: Ein Einführendes Handbuch zur Akteur-Netzwerk-Theorie,* edited by Andréa Belliger and David J. Krieger, 259–307. Zürich/ Berlin: transcript Verlag, 2006.

——. 'The More Manipulations the Better'. In *New Representation in Scientific Practice,* edited by Catelijne Coopmans, Michael Lynch, Janet Vertesi, and Steve Woolgar, 347–50. Cambridge, MA: MIT Press, 2014.

——. *Down to Earth: Politics in the New Climatic Regime*. Cambridge, UK: Polity, 2018.

Latour, Bruno, and Steve Woolgar. *Laboratory Life: The Construction of Scientific Facts*. Edited by Jonas Salk. Princeton, NJ: Princeton University Press, 1979.

——. *Laboratory Life: The Construction of Scientific Facts,* 2nd ed. 1986. https://www.jstor.org/stable/j.ctt32bbxc.

Layton-Matthews, Kate, Arpat Ozgul, and Michael Griesser. 'The Interacting Effects of Forestry and Climate Change on the Demography of a Group-Living Bird Population'. *Oecologia* 186, no. 4 (2018): 907–18. https://doi.org/10.1007/s00442-018-4100-z.

Le Guin, Ursula K. 'The Carrier Bag Theory of Fiction'. In *The Ecocriticism Reader: Landmarks in Literary Ecology,* 149–54. Athens, GA: University of Georgia Press, 1996.

Lindgren, Folke. 'Iakttagelser Rörande Lavskrikan *(Perisoreus Infaustus)* Huvudsakligen Dess Häckningsbiologi'. *Fauna och Flora* 70 (1975): 193–232.

Lowenhaupt Tsing, Anna. *The Mushroom at the End of the World: On the Possibility of Life in Capitalist Ruins*. Princeton, NJ: Princeton University Press, 2015.

Manceron, Vanessa. *Wild and Wonderful: An Ethnography of English Naturalists*. Translated by Michael Taylor. HAU, 2025. https://press.uchicago.edu/ucp/books/book/distributed/W/bo239333429.html.

Marder, Michael. *Plant-Thinking: A Philosophy of Vegetal Life*. New York: Columbia University Press, 2013.

——. 'For a Phytocentrism to Come'. *Environmental Philosophy* 11, no. 2 (2014): 237–52. https://doi.org/10.5840/envirophil20145110.

Mareis, Claudia. *Design als Wissenskultur: Interferenzen zwischen Design- und Wissensdiskursen seit 1960*. Bielefeld: Transcript, 2011.

Mareis, Claudia, Christof Windgätter, Paul Feyerabend, John Law, Knut Ebeling, Stefan Wellgraf, Anja Schwanhäußer, et al. *Wild Thing. Unordentliche Prozesse in Design und Wissenschaft*. Edited by Claudia Mareis and Christof Windgätter. Berlin: Kulturverlag Kadmos, 2019.

Margulis, Lynn. *The Symbiotic Planet: A New Look At Evolution*. New York: Phoenix, 1998.

Martin, Paul, and Patrick Bateson. *Measuring Behaviour: An Introductory Guide*. Cambridge: Cambridge University Press, 2007.

Mauss, Marcel. 'Techniques of the Body'. In *Journal de Psychologie Normal et Patholigique,* edited by Marcel Mauss, 271–93. London: Routledge, 1935.

Maye, Harun. 'Was Ist Eine Kulturtechnik?' *Zeitschrift für Medien und Kulturforschung,* no. Heft 1 (2010): 121–35.

Merz, Martina. 'Bildkomplexe als Geschichten: Naturwissenschaftler erzählen'. In *Erzählen in den Wissenschaften.* Fribourg: Academic Press Fribourg, 2009.

Merz, Martina, and Inge Hinterwaldner. 'Neue Bilder, Modelle und Simulationen: Zwischen Repräsentativität und Produktivität'. In *Handbuch Wissenschaftssoziologie,* edited by Sabine Maasen, Mario Kaiser, Martin Reinhart, and Barbara Sutter, 303–16. Wiesbaden: Springer, 2012.

Mitchell, Claudia, Linda Theron, Ann Smith, and Jean Stuart. 'Picturing Research'. In *Picturing Research,* 1–16. Rotterdam: SensePublishers, 2011. https://doi.org/10.1007/978-94-6091-596-3_1.

Morton, Timothy. *Dark Ecology: For a Logic of Future Coexistence.* New York: Columbia University Press, 2016.

Mundy, Rachel. 'Birdsong and the Image of Evolution'. *Society & Animals* 17, no. 3 (2009): 206–23. https://doi.org/10.1163/156853009X445389.

Nasim, Omar W. 'Extending the Gaze: The Temporality of Astronomical Paperwork'. *Science in Context* 26, no. 2 (2013): 247–77. https://doi.org/10.1017/S0269889713000057.

Niehaus, Amanda C. 'Tell the Stories in Your Science'. *Nature* 557 (2018): 269.

Nystrand, Magdalena, Michael Griesser, Sönke Eggers, and Jan Ekman. 'Habitat-Specific Demography and Source-Sink Dynamics in a Population of Siberian Jays'. *Journal of Animal Ecology* 79, no. 1 (2010): 266–74.

Offenhuber, Dietmar. 'Data by Proxy: Material Traces as Autographic Visualizations'. *IEEE Transactions on Visualization and Computer Graphics* 26, no. 1 (2020): 98–108. https://doi.org/10.1109/TVCG.2019.2934788.

———. 'Dis/Entangling Perspectives in Material Research'. In *Data Autographies: A Material Perspective on Data Visualization and Evidence Construction.* Online publication, Matter of Activity: Humboldt University Berlin, 2022. https://www.matters-of-activity.de/en/activities/6386/dis-entangling-perspectives-in-material-research.

O'Reilly, Karen. *Ethnographic Methods.* New York: Routledge, 2005. https://doi.org/10.4324/9780203864722.

Pink, Sarah. 'Doing Sensory Ethnography'. In *Doing Sensory Ethnography,* 7–23, London: SAGE Publications, 2009. https://doi.org/10.4135/9781446249383.n2.

Polanyi, Michael, and Amartya Sen. *The Tacit Dimension.* Chicago: University of Chicago Press, 2006.

Pötzsch, Frederik S., and Bernt Schnettler. 'Visuelles Wissen'. *Handbuch Wissenssoziologie und Wissensforschung,* 2007.

Renger, Almut-Barbara, Christoph Wulf, Jan Ole Bangen, and Henriette Hanky. 'Körperwissen: Transfer und Innovation'. *Paragrana* 25, no. 1 (2016): 13–19.

Rheinberger, Hans-Jörg. 'Experimentalsysteme: Differenz, Graphematizität, Konjunktur'. In *Experiment, Differenz, Schrift. zur Geschichte Epistemischer Dinge,* 21–32. Basilisken-Presse, 1992.

———. 'Toward a History of Epistemic Things: Synthesizing Proteins in the Test Tube'. *Historical Studies in the Physical and Biological Sciences* 27, no. 2 (1997): 325.

Rillig, Matthias C., and Karine Bonneval. 'The Artist Who Co-Authored a Paper and Expanded My Professional Network', *Nature,* 27 February (2020): 1–8.

Rouse, Joseph. 'Barad's Feminist Naturalism'. *Hypatia* 19, no. 1 (2004): 142–61. https://doi.org/10.1353/hyp.2004.0012.

Schön, Donald A. *The Reflective Practitioner: How Professionals Think in Action.* Edited by Donald A. Schön. New York: Basic Books, 1983.

Sheldrake, Merlin. *Entangled Life: How Fungi Make Our Worlds, Change Our Minds, and Shape Our Futures.* London: The Bodley Head, 2020.

Simon, Herbert A. *The Sciences of the Artificial.* Edited by Herbert A. Simon. Vol. 33. Cambridge, MA: MIT Press, 1997.

Soga, Masashi, and Kevin J. Gaston. 'Extinction of Experience among Ecologists'. *Trends in Ecology & Evolution* 40, no. 3 (March 2025): 212–15. https://doi.org/10.1016/j.tree.2024.12.010.

Taussig, Michael. *Mimesis and Alterity: A Particular History of the Senses.* New York: Routledge, 1993.

———. 'What Do Drawings Want?' *Culture, Theory and Critique* 50, no. 2–3 (2009): 263–74.

———. *I Swear I Saw This: Drawings in Fieldwork Notebooks, Namely My Own.* Chicago: University of Chicago Press, 2011.

Thyssen, Geert, and Ian Grosvenor. 'Learning to Make Sense: Interdisciplinary Perspectives on Sensory Education and Embodied Enculturation'. *The Senses and Society* 14, no. 2 (2019): 119–30. https://doi.org/10.1080/17458927.2019.1621487.

Thyssen, Geert, and Karin Priem. 'Mobilising Meaning: Multimodality, Translocation, Technology and Heritage'. *Paedagogica Historica* 49, no. 6 (2013): 735–44. https://doi.org/10.1080/00309230.2013.848912.

Torp, Eivind. 'The Legal Basis of Sami Reindeer Herding Rights in Sweden'. *Arctic Review on Law and Politics* 1, no. 4 (2013): 43–61.

Tsing, Anna. 'When the Things We Study Respond to Each Other: Tools for Unpacking "the Material"'. In *Anthropos and the Material,* edited by Penny Harvey, Christian Krohn-Hansen, and Knut G. Nustad, 221–43. Durham, NC: Duke University Press, 2019.

Tsing, Anna Lowenhaupt, Andrew S. Mathews, and Nils Bubandt. 'Patchy Anthropocene: Landscape Structure, Multispecies History, and the Retooling of Anthropology: An Introduction to Supplement 20'. *Current Anthropology* 60, no. S20 (2019): S186–97. https://doi.org/10.1086/703391.

Turner, Sally, and Joseph Leydon. 'Improving Geographic Literacy among First-Year Undergraduate Students: Testing the Effectiveness of Online Quizzes'. *Journal of Geography* 111, no. 2 (2012): 54–65. https://doi.org/10.1080/00221341.2011.583263.

Van Dooren, Thom. *Flight Ways: Life and Loss at the Edge of Extinction.* Critical Perspectives on Animals: Theory, Culture, Science, and Law. New York: Columbia University Press, 2016.

Voss, Julia. *Darwins Bilder: Ansichten der Evolutionstheorie 1837–1874.* Berlin: Fischer Taschenbuch Verlag, 2007.

Whitehouse, Andrew. 'Listening to Birds in the Anthropocene: The Anxious Semiotics of Sound in a Human-Dominated World'. *Environmental Humanities* 6 (2015): 53–71.

Willerslev, Rane. 'Not Animal, Not Not-Animal: Hunting, Imitation and Empathetic Knowledge among the Siberian Yukaghirs'. *Journal of the Royal Anthropological Institute* 10, no. 3 (2004): 629–52.

Willkomm, Judith. 'Mediatisierte Sinne und die Eigensinnigkeit der Medien. Für eine Medientheoretische Sensibilisierung der Sinnlichen Ethnographie'. In *Ethnographien der Sinne*, edited by Lydia Maria Arantes and Elisa Rieger, 39–56. Bielefeld: Transcipt Verlag, 2014.

Image Reference

The photographs and scans were produced by the author. The originals were provided by the PI of the case study, Michael Griesser, and are used with his consent.